LOOKING AT
Landforms

MICHAEL DAY
SUSAN DAY
IAIN MEYER

Nelson

Acknowledgements

The authors and publishers are grateful to the following for permission to use their photographs:

Aerofilms pp. 59, 61, 85, 88; Associated Press pp. 14, 18, 83; Geoffrey Attridge p. 36; Barnaby's Picture Library pp. 29, 80; BBC Hulton Picture Library pp. 6, 34; Calorific p.91; Celtic Picture Agency pp. 42, 76; Daily Telegraph Colour Library p. 9; Foreign Language Press p. 55; Forestry Commission p. 32; Robert Harding Picture Library p. 24; Alan Hutchinson Library pp. 40, 45, 49; P.G. Laws p. 22; R.N. Partridge p. 86; Chris Ridgers Photography p. 36; J.L. Rock p. 81; Thames Water Board p. 39; Elsie Timby Collection for the Society for Cultural Relations with the USSR p. 78; Zefa cover photograph and pp. 5, 12, 20, 47, 53, 56, 64, 71, 73, 92.

We are indebted to the following for permission to reproduce news extracts:

Financial Times p. 19; *The Guardian* p. 81; *New Scientist* p.79; *The Strand Magazine* p.7; *The Sunday Times* pp. 6, 25; *The Times* p. 37.

Cover Photograph: Rice growing in the tropical humid climate of the Philippines. The terraces control the flow of water and help to prevent soil erosion on the valley sides.

Illustrated by Richard Bonson, Jane Cheswright and Howard Prescott.

Thomas Nelson and Sons Ltd
Nelson House Mayfield Road
Walton-on-Thames Surrey
KT12 5PL UK

51 York Place
Edinburgh
EH1 3JD UK

Thomas Nelson (Hong Kong) Ltd
Toppan Building 10/F
22A Westlands Road
Quarry Bay Hong Kong

Thomas Nelson Australia
102 Dodds Street
South Melbourne
Victoria 3205 Australia

Nelson Canada
1120 Birchmount Road
Scarborough Ontario
MIK 5G4 Canada

First published by Thomas Nelson and Sons Ltd 1987

ISBN 0-17-444067-7
NPN 9 8 7 6 5 4 3

Printed in Great Britain

Contents

CHAPTER 1: *Restless earth*

Violent geological events	6
Colliding continents	8
Plate boundaries	10

CHAPTER 2: *Mountains, earthquakes and volcanoes*

Mountain building	12
Case study: The Mexican earthquake	14
Earthquakes	15
Volcanoes	16
Case study: The Nevado del Ruiz, Colombia	18
Magma intrusions	20
Hydrothermal activity	21

CHAPTER 3: *Weathering*

Weathering processes	22
Acid rain – weathering by pollution	24
Limestone scenery – a weathered landscape	26

CHAPTER 4: *Hydrology*

Drainage basins	28
The drainage basin system	30
Controlling the drainage basin system	32
Flooding and flood control	34
Case study: The Brent flood	36
Water supply	38
Case study: The Kano River Irrigation Project	40

CHAPTER 5: *Rivers and river valleys*

The river valley system	42
The effect of rainfall on slopes	44
The effect of gravity on slopes	46
Case study: Soil erosion in Nepal	48
The power of rivers	50
Erosion, transport and deposition in the river	51
Case study: China's Sorrow – the Hwang Ho or Yellow River	54
The long profile of a river	56
Meanders	58
Floodplains	60
Terraces	63
Deltas	63

Contents

CHAPTER 6: *Ice sheets and glaciers*

Ice sheets today	64
Ice ages in the past	65
The valley glacier system	66
Glacier flow	68
Valley glacier erosion	69
Valley glacier deposition	72
Meltwater	74
Case study: The Nant Ffrancon – a glaciated valley	76
At the edge of the ice sheet	78

CHAPTER 7: *Coasts*

Waves	80
Tides and tidal surges	82
Coastal erosion	84
Case study: The erosion of Barton Cliffs	86
Coastal deposition	87
Case study: Deposition in the Dovey estuary	88
Sea-level change	90
The crowded coast	90
Case study: Venice – coastal mismanagement	91

CHAPTER 8: *Deserts – The great arid lands*

Shaping the desert	92
Desertification – the expanding deserts	94
Case study: The Sahel	95
Index	96

About this book

Mule train climbing up from Baudada, deep in the Himalayan mountains of Nepal. Has the landscape influenced human activity? Has human activity influenced the landscape?

This book covers those aspects of the study of landforms and related topics which are required by the current syllabuses for GCSE examinations in Geography.

The authors' approach reflects recent thinking in the subject. As such, this book has the following aims:

● To introduce pupils to the wonder that exists in the natural world, for example, the fact that whole continents are on the move, that much of the earth was once covered by great thicknesses of ice, and that it is possible to see evidence of such facts in the landscape of our own country.

● To show pupils the direct but complex relationship between human activity and the physical environment. This relationship can be terrifyingly destructive as in the case of a natural hazard such as the Mexican earthquake of 1985, or when our interference with the natural environment leads to major problems such as dramatically increased weathering rates induced by acid rainfall. There are also examples of where people have learnt to control their physical environment, such as the engineering achievement of building new road and rail routes across the permafrost of Siberia.

● To demonstrate new ideas for understanding the complex interactions between processes in the natural world – for example, the use of simple flow and system diagrams to appreciate the formation of a river valley.

● To involve pupils in applying their new knowledge to the production of solutions to common environmental problems, such as the control of serious urban flooding in the River Brent drainage basin of North London.

● To help pupils appreciate the need to value the beauty of the natural world, such as the dramatic glacial features to be found along the Nant Ffrancon valley in Wales, and to understand differing attitudes towards environmental issues.

● To help pupils to appreciate the need to manage resources provided by the natural world.

● To make learning fun, by involving the pupils in meaningful exercises, which help to develop skills and encourage the discovery of new ideas.

Inevitably, the nature of a textbook means that the authors have had to be selective. It is expected that teachers will want to use the book in order to develop a core unit in this topic, and then go beyond this with resources of their own. To help with this organisation the book has been designed around a series of double page spreads, each of which looks at one specific topic. The text and exercises have been carefully graded in difficulty, much more being expected of the pupils as the book develops.

The authors believe that true learning can only take place when the pupils find the work fun, meaningful and stretching. It is their hope that this book will form the foundation for such a course for any GCSE landform unit.

ONE

Restless earth

Violent geological events

What is it like to be caught up in an earthquake? When a volcano erupts, what happens to the people living near it? Read the two extracts on this page to find out.

◀ St Pierrè, Martinique, after the volcanic blast–30 000 dead

An earthquake in Guatemala City

Suddenly there was a tremendous roar, as though a lot of railway trains were rattling and rumbling through the darkness of the room. A fine cloud of plaster began to fall from the ceiling. The outside wall yawned open and disappeared into the night, and lathes and beams and great chunks of ceiling began to fall all about me. I turned over on to my stomach, prepared at any moment to be pitched face first into the void.

Gradually the tumult began to die down. The air was thick with dust, and I and my bed were buried in masonry and plaster. In inky blackness I fumbled about on my bedside table where I had a torch; the table was now deep in debris. I leapt out of bed, grabbed my shoes and tried to pull them on but they were full of rubble.

Thor Heyerdahl,
The Sunday Times, February 1976.

Clearly, living through an earthquake or a volcanic eruption is a shattering experience. It seems to most people that they happen without any warning. But there may be a pattern to this activity. Complete exercise 1 to find out if there is.

A volcanic explosion on Martinique

Hot ashes fell thick at first. They were soon followed by a rain of small, hot stones, ranging all the way from the size of shot to pigeon's eggs. These would drop into the water around our ship with a hissing sound. After the stones came a rain of hot mud and lava, of the consistency of very thin cement. Wherever it fell it formed a coating, clinging like glue, so that those who wore no caps it coated, making a complete cement mask right over their heads.

I snatched a tarpaulin cover off one of the ventilators and jammed it down over my head and neck, looking out through the opening. This saved me, but even so my beard, face, nostrils and eyes were so filled with the stuff that every few seconds I had to break it out of my eyes in order to see. I remember that Charles Thompson had his head so weighted down with the stuff that he seemed to feel giddy and was almost falling. When he asked me to break the casting off his head I was afraid it would scalp him.

Chief Officer Ellery Scott,
The Strand Magazine, September 1902.

Exercise 1.1

1 Look at the following two statements. Say whether you think they are true or false. Give the reasons for your answers.
'Earthquakes and volcanic eruptions are rare events.'
'Very few people live in areas prone to earthquakes and volcanic eruptions.'

2 Table 1.1 shows where in the world major earthquakes and volcanic eruptions have occurred.
a) Plot the locations on a copy of a world map. Use small dots to show the earthquakes and triangles to show the volcanoes.
b) Briefly describe the pattern of earthquakes and volcanoes shown on your map.
c) What do you think this tells you about the earth's surface?

3 From what you have just learnt in question 2, would you change any of your answers to question 1? Why?

▶ Table 1.1

Some major earthquakes

Plot these using ●

Location	Year	Place	Comments
California	1906	San Francisco	700 killed, city destroyed
Morocco	1960	Agadir, near Marrakesh	14 000 killed
Yugoslavia	1963	Skopje	1200 killed
Alaska	1964	Anchorage	Major earthquake, 130 killed
Peru	1970	Huanaco, central Peru	30 000 killed
Guatemala	1976	Guatemala City	22 000 killed
Turkey	1976	Sivas	5000 killed, remote area
Italy	1980	Southern Appenine Mts	3000 killed
Mexico	1985	Mexico City	4000 killed, city badly damaged

Some major volcanic eruptions

Plot these using ▲

Location	Year	Place	Comments
Indonesia	1815	Tamboro Island, near Flores	12 000 killed
Indonesia	1883	Krakatoa Island, West Java	36 419 killed, island exploded with the force of 26 hydrogen bombs
Lesser Antilles	1902	Mount Pelée, Martinique	30 000 killed instantly
Indonesia	1963	Agung Mountain, Bali Island, West Java	1584 killed, created major dust cloud
USA	1980	Mt St Helens, near Tacoma, Washington State	61 killed, major explosion and dust cloud
Mexico	1982	El Chichón, Yucatan Peninsula	Major explosion, dust cloud drift around the earth
Sicily	1983	Mt Etna	Eruption threatened towns and villages with lava
Colombia	1985	Nevado del Ruiz, 128 km west of Bogotá	22 000 people killed by major mud flows

Use the index of your atlas to find these places.

Colliding continents

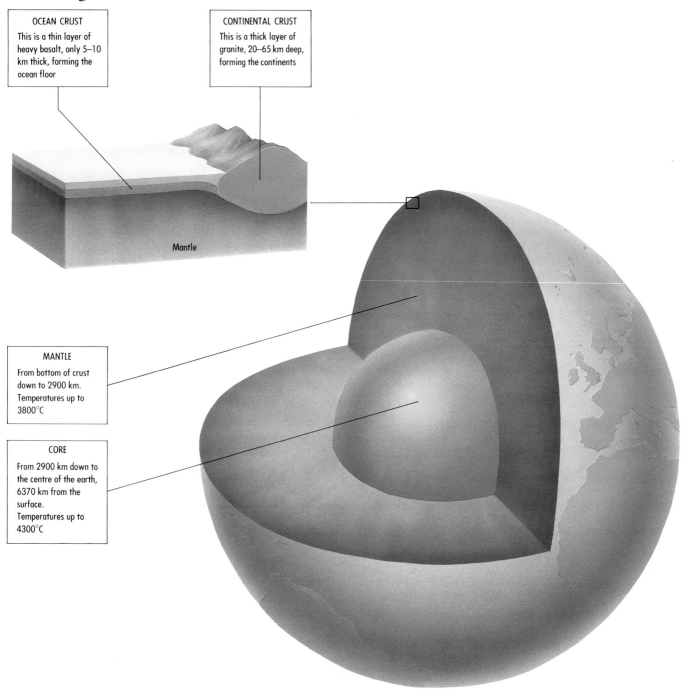

OCEAN CRUST

This is a thin layer of heavy basalt, only 5–10 km thick, forming the ocean floor

CONTINENTAL CRUST

This is a thick layer of granite, 20–65 km deep, forming the continents

Mantle

MANTLE

From bottom of crust down to 2900 km. Temperatures up to 3800°C

CORE

From 2900 km down to the centre of the earth, 6370 km from the surface. Temperatures up to 4300°C

▲ Figure 1.1 The structure of the earth

Earthquakes and volcanoes occur in definite patterns because giant sections of the earth's surface, called **plates**, are always on the move. How were these plates formed, and why do they move?

Figure 1.1 shows a cross section through the earth. Notice that the outside layer of our planet is called the earth's crust. The crust is made up of two main types of rock – **continental crust** and **ocean crust**.

Continental crust is a relatively light rock. It is broken up into gigantic pieces which float like huge icebergs on top of the **mantle** (look at Figure 1.1 again). These pieces form the continents. Surrounding the continents is the much heavier and thinner ocean crust.

Under the earth's crust is the mantle. Near to the surface, the mantle is still solid, and glued to the crust. Below 250 kilometres, however, intense heat and pressure has turned the mantle into a 'slushy' layer. Some parts of the slushy mantle are hotter than others. These differences in temperature set up heat or **convection currents**, as you can see in Figure 1.2.

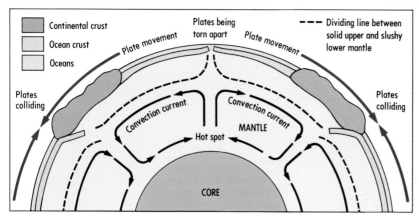

▲ Figure 1.2 Convection currents in the lower mantle split the earth's crust into a series of plates

Although the convection currents can only move very slowly (a few centimetres per year), their effect is very dramatic. As the convection currents move in different directions, they cause friction at the earth's surface. This tears the surface apart, as shown in Figure 1.2. So, the earth's surface has been split up into a number of huge rafts or **plates**, each one moving in the direction of its own convection current. This means that over millions of years the continents have been constantly changing their positions. This process is called **continental drift**.

Sample number	Distance from the Mid-Atlantic ridge (km)	Age of rock sample (millions of years)
1	200	10
2	420	24
3	520	27
4	750	32
5	860	40
6	**1000**	48
7	1300	66
8	1680	78

▲ Table 1.2

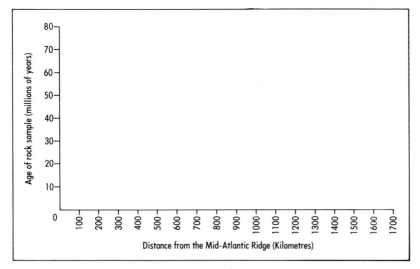

▲ Figure 1.3

Exercise 1.2

In 1968, scientists wanted to find out the age of the ocean crust under the Atlantic ocean. The American drilling ship *Glomar Challenger* drilled a series of bore holes across various parts of the Atlantic Ocean floor. Samples of rock were taken from each hole and tested. Table 1.2 shows the results of one survey line.

1 Copy Figure 1.3 onto graph paper and plot the information shown in Table 1.2.
2 The Mid-Atlantic Ridge is thought to be a place where the ocean floor is splitting apart, and new rock is being formed. Write a short paragraph explaining why your graph confirms this idea.

◀ Earth from Space – not as quiet as it seems

Plate boundaries

The place where two plates touch each other is called a **plate boundary**. There are three main types of plate boundary. Each affects the earth's surface in a different way.

Example: The Western Edge of South America

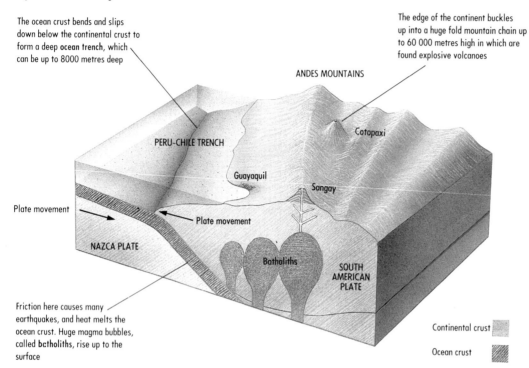

The ocean crust bends and slips down below the continental crust to form a deep **ocean trench**, which can be up to 8000 metres deep

The edge of the continent buckles up into a huge fold mountain chain up to 60 000 metres high in which are found explosive volcanoes

ANDES MOUNTAINS

Cotopaxi

PERU-CHILE TRENCH

Guayaquil

Sangay

Plate movement

Plate movement

NAZCA PLATE

Batholiths

SOUTH AMERICAN PLATE

Friction here causes many earthquakes, and heat melts the ocean crust. Huge magma bubbles, called **batholiths**, rise up to the surface

Continental crust

Ocean crust

◀ Figure 1.4
Destructive plate boundary
This is where two plates are colliding together. Gigantic forces buckle the earth's surface into huge mountains and cause volcanoes to occur.

Example: The Mid-Atlantic Ridge

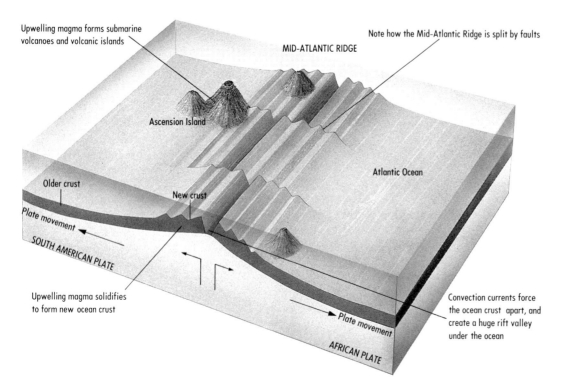

Upwelling magma forms submarine volcanoes and volcanic islands

MID-ATLANTIC RIDGE

Note how the Mid-Atlantic Ridge is split by faults

Ascension Island

Atlantic Ocean

Older crust

New crust

Plate movement

SOUTH AMERICAN PLATE

Upwelling magma solidifies to form new ocean crust

Plate movement

Convection currents force the ocean crust apart, and create a huge rift valley under the ocean

AFRICAN PLATE

◀ Figure 1.5
Constructive plate boundary
This is where two plates are being wrenched apart, creating a rift valley, where new ocean crust is formed.

Example: The San Andreas Fault

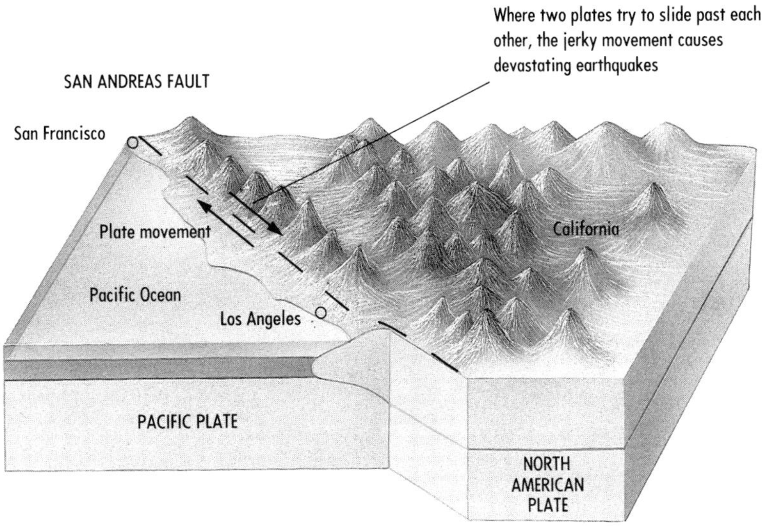

SAN ANDREAS FAULT

Where two plates try to slide past each other, the jerky movement causes devastating earthquakes

San Francisco

Plate movement

California

Pacific Ocean

Los Angeles

PACIFIC PLATE

NORTH AMERICAN PLATE

▶ **Figure 1.6**
Slipping plate boundary When two plates try to slip past each other, tremendous friction makes the movement jerky, causing earthquakes.

Exercise 1.3

1 **a)** Figure 1.7 shows the major world plates, labelled A–G. Match each plate to its name, shown in the list below.
African plate
Antarctic plate
Indo–Australian plate
North American plate
Pacific plate
South American plate
Nazca plate
b) On a copy of a world map, shade in the following fold mountain ranges.
Rocky Mountains
Andes
Pyrenees

Alps
Himalayan Mountains

2 **a)** On which type of plate boundary do the following places lie?
San Francisco
Mediterranean Sea
Iceland
b) Using sketch diagrams to help you, explain the following three statements:
'San Francisco is not a safe place to live.'
'The Mediterranean Sea will no longer exist 100 million years from now.'
'Iceland is a living laboratory for geologists, with a growth rate of two centimetres per year.'

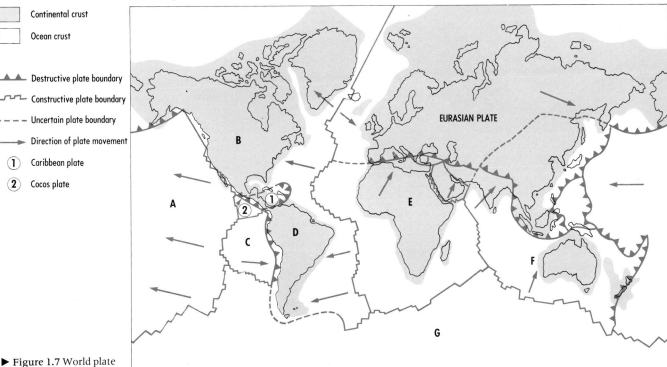

Continental crust

Ocean crust

▲▲▲ Destructive plate boundary

ⵜⵍⵜ Constructive plate boundary

--- Uncertain plate boundary

→ Direction of plate movement

① Caribbean plate

② Cocos plate

EURASIAN PLATE

A   B   C   D   E   F   G   ①   ②

▶ **Figure 1.7** World plate boundaries

TWO

Mountains, earthquakes and volcanoes
Mountain building

Look at the photograph of Mount McKinley. Most of the world's great mountains have been formed at plate boundaries. Two important processes occurring during mountain building are folding and faulting.

Folding

When two plates collide, the rock layers on the earth's surface are put under tremendous pressure. Sometimes they bend under the pressure and **fold**. Figure 2.1 illustrates the different types of fold that can be produced.

An example of folding is shown in Figure 2.2. This occurred 25 million years ago when the African plate collided with the Eurasian plate. A flat chalk plain in southern Britain was folded up into a series of hills (**anticlines**), and valleys (**synclines**). As often happens, stress caused the Weald anticline to crack along its summit. Gradually, rivers eroded (wore away) the rock along the line of weakness. This left chalk **escarpments** on either side which today form the North and

▲ Mount McKinley, Alaska, USA

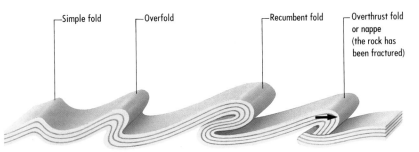

▲ Figure 2.1 Types of folding

South Downs. The syncline was later filled by river and sea mud, to form the London Basin clays.

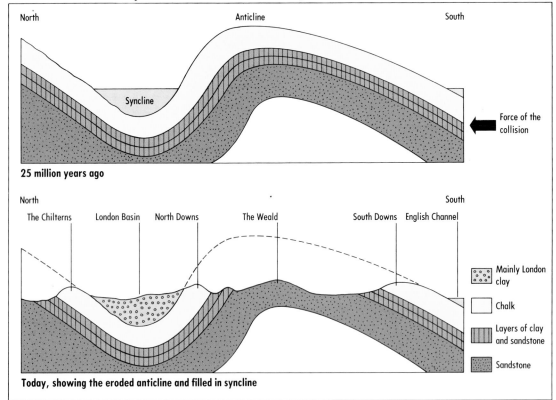

◀ Figure 2.2 A cross section through South East England

Faulting

Faulting occurs when stress at a plate boundary makes the rock tear, rather than fold. The line of the tear, or break, is known as a **fault**. Figure 2.3 shows the different types of fault that can occur.

Some of the world's most spectacular faults occur when continental crust is split open at a constructive plate boundary. (If you need to remind yourself about constructive plate boundaries, look at page 10.) Convection currents in the lower mantle push the crust up into a vast dome. The dome's surface splits into a Y-shaped pattern of **rift valleys**. On either side of the valleys are block mountains, or **horsts**. Volcanoes occur along the line of the faults (Figure 2.4.). One example of this is the East African rift valley system, which stretches for over 4800 kilometres. You can see this in your atlas.

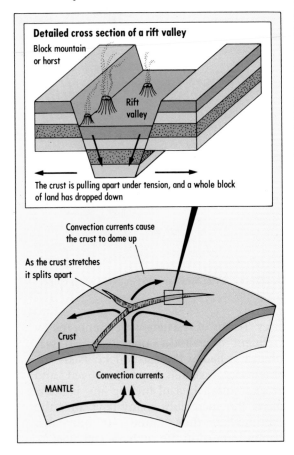

▶ Figure 2.4 The formation of a rift valley

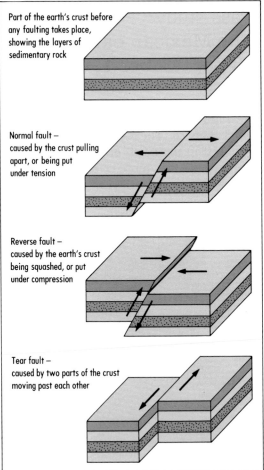

▲ **Figure 2.3** Types of fault

Exercise 2.1

1   Look at Figure 2.5. Name features A-E from the following list: syncline, anticline, fault, rift valley, escarpment.

2   a) Find the cross section line on an atlas map of Scotland. Name the two hill ranges forming the lava escarpments either side of the rift valley.
    b) Describe what has happened to the rift valley since its formation.
    c) Name the valley formed by the syncline.

3   Use your atlas to help you describe the effect of geology upon settlement and communication.

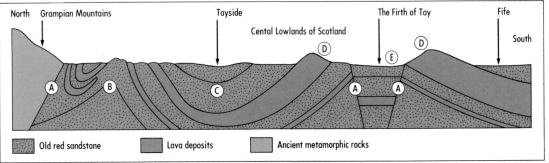

▶ Figure 2.5 A simplified geological cross section of the Tay Region

Case study:
The Mexican earthquake, 19 September 1985

Early one Thursday morning, at 7.20 a.m., the third largest earthquake this century struck Mexico. The Cocos plate (a small plate just to the north of the Nazca plate) had slipped a few more centimetres under the Caribbean plate. Over 7000 people died.

▲ Rescue workers digging frantically for survivors

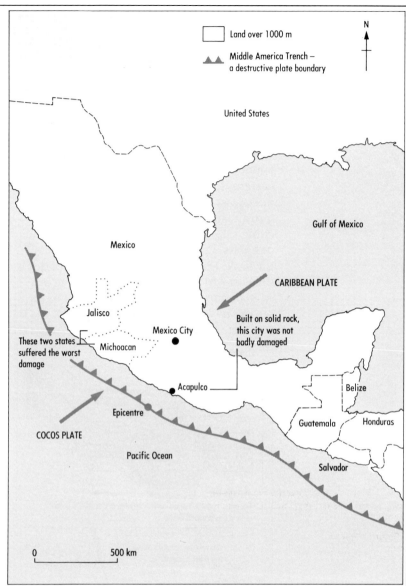

▲ **Figure 2.6** The Mexican earthquake, 19 September 1985

The earthquake occurred beneath the floor of the Pacific Ocean (Figure 2.6). A 30 metre high tsunami (a huge earthquake generated seawave) swamped two cargo ships and fifteen fishing boats. Most of the world's attention, however, was focused on the plight of the capital – Mexico City, which has a population of eighteen million. The city is built on deep layers of sand and mud. During the earthquake, this unstable foundation shook like jelly. In some places, pockets of air were released from the ground, causing massive subsidence.

The result was devastating. Hundreds of buildings collapsed, including those supposed to be earthquake-proof. Water mains were burst open, sewage systems ruptured, and broken gas mains started thousands of fires. Teams of rescuers were hampered by frequent aftershocks, one of which was almost as bad as the earthquake itself. The three city hospitals were destroyed, and rescuers despaired of finding any patients alive. But incredibly, many days after the earthquake, they found 40 new-born babies trapped in the rubble – still alive.

The cost of looking after thousands of homeless people and rebuilding towns and cities is considerable. The Mexican government already had a huge foreign aid debt of $100 billion. They found it difficult to borrow yet more money, though both the World Bank and the IMF did lend them large sums. There is much concern that Mexico may never be able to afford to repay these mounting debts.

Earthquakes

Major earthquakes normally occur either at slipping or destructive plate boundaries. Here, the crust is subjected to enormous stress as the plates try to move in opposite directions. The rocks are forced to bend, building up huge quantities of stored energy like a watch spring. Eventually, the stress is so great that the rocks break and snap back along a fault line. The energy is released in **seismic waves** which make the ground shake and produce an earthquake. This is shown diagrammatically in Figure 2.7.

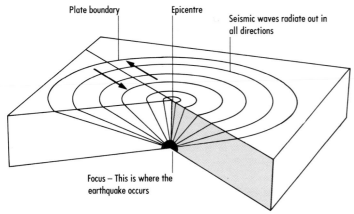

Plate boundary   Epicentre   Seismic waves radiate out in all directions

Focus – This is where the earthquake occurs

The most destructive earthquakes are those with a shallow focus, occurring at slipping plate boundaries

▲ **Figure 2.7** Seismic waves

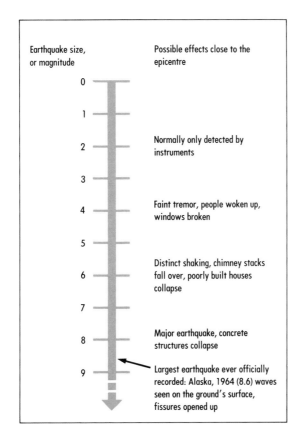

Earthquake size, or magnitude	Possible effects close to the epicentre
0	
1	
2	Normally only detected by instruments
3	
4	Faint tremor, people woken up, windows broken
5	
6	Distinct shaking, chimney stacks fall over, poorly built houses collapse
7	
8	Major earthquake, concrete structures collapse
9	Largest earthquake ever officially recorded: Alaska, 1964 (8.6) waves seen on the ground's surface, fissures opened up

▶ **Figure 2.8** The Richter scale

Primary and secondary effects

A primary earthquake effect is any damage caused as a direct result of the ground shaking. These effects include the collapse of buildings and bridges, and the rupturing of gas mains.

The primary earthquake effects cause other damage, called secondary earthquake effects. These include fires, set off by fallen stoves and broken electricity cables, disease, caused by poor sanitation and decomposing bodies, and **tsunamis**, which are large sea waves caused by a sudden movement of the sea floor.

In Japan, where earthquakes are quite common, the government has made great efforts to reduce the effects of earthquakes. Flimsy wood and brick houses have been replaced by low sturdy concrete buildings which can better withstand shaking. Narrow streets have been replaced by wide avenues that cannot be blocked by rubble, or crossed easily by fire. People in schools, offices and factories, etc. have regular earthquake drills in which they practise evacuating the building quickly.

Exercise 2.2

1 Copy out and complete the following paragraph, using the words supplied in the list below.

   'Earthquakes occur when two ____ try to slip past each other in different directions. This normally happens at a ____ or a ____ plate boundary. The rocks are put under a great deal of ____ and start to ____. Eventually, when the stress is too great, the rocks suddenly snap back along a ____ ____. Energy in the form of ____ ____ radiates out in all directions, causing the ground to ____.

   bend, shake, seismic waves, line, slipping, fault, plates, stress, destructive

2 Working in pairs, list the primary and secondary earthquake effects that occurred in Mexico in September 1985.

3 A huge earthquake occurred in Alaska in 1964. Use your school library to find out about this event. Write an account of the earthquake, explaining how it was caused, and what effects it had.

15

Volcanoes

There are two main types of volcano. Explosive and steep-sided acid lava volcanoes are found at destructive plate boundaries, and more gentle-sided basic lava volcanoes are found at constructive plate boundaries.

An acid lava volcano

Acid lava volcanoes are formed at destructive plate boundaries. Basalt ocean crust and marine sediments melt as they descend into the mantle. This forms an acid magma which is viscous (does not flow easily). It rises towards the surface as a series of huge domes, or batholiths. Pipes lead off from the batholith to form volcanoes.

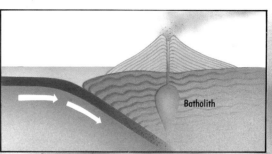

Batholith

Ash is the most common material ejected by a violent eruption along with rocks and blobs of lava. Sometimes hot ash clouds flow down the sides of the cone at 200 km/h. These are called nuées ardentes (fiery clouds), and cause huge damage.

Since the lava cools quickly, it often blocks the pipe. This usually causes a massive explosion as pressure builds up inside the volcano. Generally the pressure is so great that it blows the top of the volcano off. The result is a huge crater often over 25 km across, called a caldera.

Viscous lava is extruded out of the crater like thick toothpaste, and cools rapidly to form a steep sided cone. The cone is built up from layers of ash and lava as the volcano erupts.

Sometimes, rising magma melts snow and ice on the cone's summit. This causes devastating mud flows which travel down the slope at up to 90 km/h.

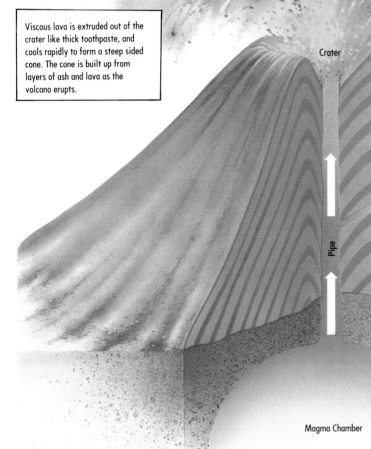

Crater

Conelet

Pipe

Pipe

Magma Chamber

Magma is molten rock. When it reaches the surface, hot gases and superheated steam violently escape out of the mixture, which then becomes lava.

Example: Mt Etna, Sicily

A basic lava volcano

Basic lava volcanoes are formed at constructive plate boundaries. The magma comes directly from upwelling molten rock, which is runny and basic (basic means neutral). The runny lava allows gentle eruptions to take place.

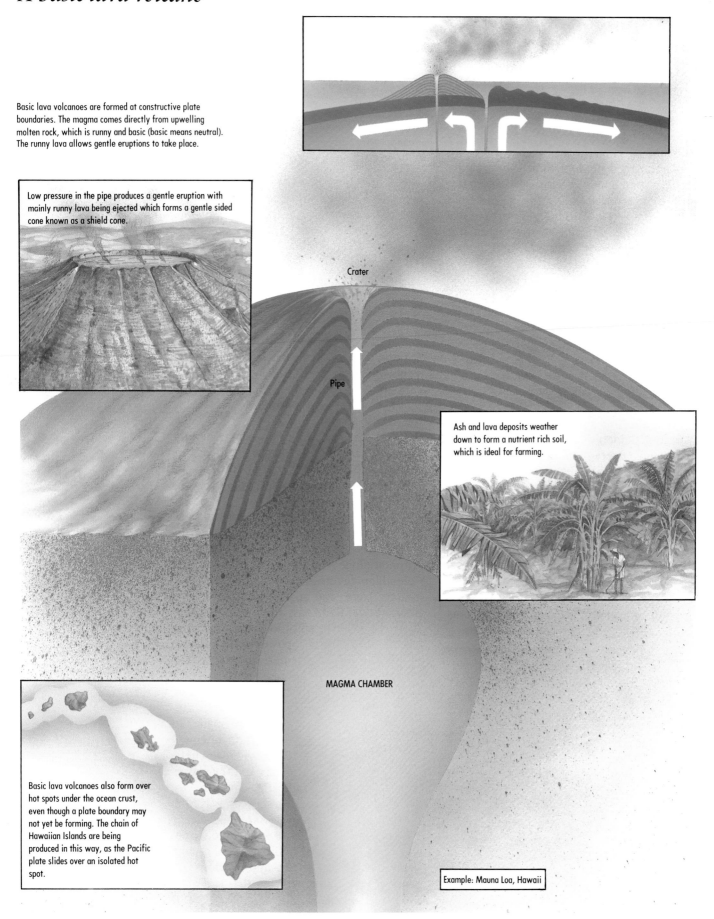

Low pressure in the pipe produces a gentle eruption with mainly runny lava being ejected which forms a gentle sided cone known as a shield cone.

Crater

Pipe

Ash and lava deposits weather down to form a nutrient rich soil, which is ideal for farming.

MAGMA CHAMBER

Basic lava volcanoes also form over hot spots under the ocean crust, even though a plate boundary may not yet be forming. The chain of Hawaiian Islands are being produced in this way, as the Pacific plate slides over an isolated hot spot.

Example: Mauna Loa, Hawaii

Case study:
The eruption of the Nevado del Ruiz, Colombia, 14 November 1985

One morning in mid November 1985, the whole world woke up to TV news pictures showing the plight of twelve year old Omayra Sanchez. She had become trapped in deep mud which had cascaded down the side of a volcano called the Nevado del Ruiz, completely burying towns and villages. Despite her injuries, she smiled and joked as rescuers worked frantically to free her. Tragically, she died from a major wound just before they could do so.

Nevado del Ruiz means 'The Sleeping Lion'. It is a snow-capped volcanic cone high up in the Andes Mountains, 128 kilometres west of Bogotá (Figure 2.9). A beautiful mountain, it is famous for its winter sport activities. Its lower slopes are covered by rich coffee farms and many large towns and villages.

At 2 a.m. on a Thursday morning, the volcano suddenly erupted with a huge explosion. There had been some earlier warnings that this might happen, but the local farmers were reluctant to leave their prosperous farms. At first there was no lava flow, but rocks were hurled out, and a dense ash and sulphur cloud rose 8000 metres into the air. The cloud drifted as far as the Venezuelan border 500 kilometres away. It covered much of the surrounding countryside with a thick layer of ash. Aircraft and motor vehicles had problems as the ash clogged their air intake pipes.

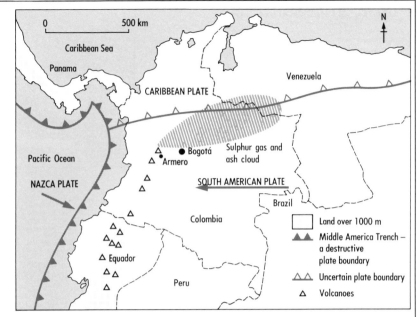

▲ Figure 2.9 The eruption of Nevado del Ruiz, 14 November 1985

The ash was not the biggest danger. The heat of the eruption was rapidly melting ice and snow on the summit of the volcano. Floodwater rushed down the local rivers, sweeping away houses and bridges. The eruption had also dislodged ancient layers of volcanic dust. This dust combined with the flood water to form an unstable volcanic mud which began to flow rapidly down the mountain. It reached speeds of over 100 kph, and one vast mudflow swept down the Languillina valley, engulfing the town of Armero.

The scale of the disaster was enormous. The death toll of 22 000 people from the mud flows made it the worst volcanic eruption since Mount Pelée in 1902. The Colombian government declared the area a military zone, and appealed for help. The International Red Cross and United Nations Disaster Relief Organisation, learning from previous disasters, co-ordinated their efforts, and flew in medical supplies, tents, digging equipment and personnel.

However, even these well-organised rescue attempts were hampered. The thin mountain air and thick dust made it difficult for rescue helicopters to operate. Destroyed bridges made it hard for rescuers to reach the worst hit areas, and there was a shortage of hand radios for communication. One of the best resources was the spirit of the Colombian people themselves. This was demonstrated to the whole world when TV pictures were broadcast showing rescuers hard at work, and millions of people were saddened by the story of Omayra Sanchez.

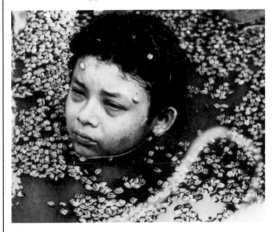

▲ Omayra Sanchez, trapped in volcanic mud and water

1 a) Read very carefully pages 16 to 17, and then copy out Table 2.1.

b) Working in pairs, complete the table by writing in the correct descriptions from the list below:

Gentle sided shield shape
Often a very violent and explosive eruption
Steep sided cone shape
At destructive plate boundaries
Directly from upwelling molten rock
Mainly runny lava, which flows rapidly away from the crater
At constructive plate boundaries
Ash, hot gases, lava bombs, and viscous lava that cools rapidly
Usually comparatively gentle eruption
From the melting of basalt ocean crust and marine sediments at a subduction zone

Characteristic	Acid lava volcano	Basic lava volcano
Shape		
Type of plate boundary where it is found		
Origin of magma		
Violence of eruption		
Materials ejected		
Example		

▶ **Table 2.1** The characteristics of acid and basic lava volcanoes

2 a) Still working in pairs, decide what type of volcano is the Nevado del Ruiz.

b) Using the table you completed in question 1, list the evidence that enabled you to identify the Nevado del Ruiz as this type of volcano.

3 Read the article about the St Vincent volcano.

a) Find St Vincent island in your atlas, and draw a sketch map to show its location in relation to Latin America and the West Indies.

b) Find four points mentioned in the article which show the serious effects of La Soufriére upon the people and economy of St Vincent.

4 Television means that the whole world quickly learns about a major tragedy such as a volcanic eruption. People soon forget how devastating such a disaster can be for people living in the area long after the event. For either the Colombian or the St Vincent example, write a newspaper story titled 'One Year On'. In your story, imagine that you have returned to the area, and that you have been asked to record the way in which the local people, the government and the economy of the country are still struggling to overcome the problems caused by the eruption.

In the Shadow of a Volcano

The Soufriere volcano on St Vincent, which rumbled into activity on Good Friday, has now been relatively quiet for three weeks. However, the Government of St Vincent, one of several poor and underdeveloped former British colonies in the Windward and Leeward Islands, will be grappling with the after-effects for some time yet. Although there is no loss of life, it has completely disrupted the social and economic life of St Vincent's 110 000 inhabitants.

The volcano's 1.5 kilometre wide crater is at the northern end of the island. By Easter Monday the Government had evacuated almost 20 000 residents of seven towns and villages in danger areas. This placed an enormous burden on a Government which had budgeted for a total expenditure of about £6.5 million for the current year. It estimates that it is now spending over £15,000 daily to keep the operation going.

There has been an immediate response from Britain, Canada and the US, and several neighbouring Caribbean states. Food, clothing and medical necessities have been rushed in. But such assistance, welcome as it is, barely scratches the surface. Most of those dislocated were small subsistence farmers. Their homes and land are now covered by choking volcanic dust, as deep as 1.5 metres in some places.

The hardest blow has been to the vital banana industry, St Vincent's principal export crop and largest single employer. Plantations in the immediate vicinity of Soufriére were devastated but in other parts of the island the fallout of ash and dust scorched the skin of the fruit. In common with other Windward Islands, St Vincent sells almost its total production to Geest Industries of Spalding, Lincolnshire. It was anticipated that exports would be worth over £4 million this year, but weekly shipments have dropped by 50 per cent.

The disaster will also have political effects. Along with others in the Windward and Leeward groups, St Vincent had indicated it would press for full independence from Britain next year. Now the plans of the Premier and his Government will have to be altered. Independence will surely be a contradiction for an island which is heavily dependent on all the outside help it can get.

Tony Cozier
Financial Times, 18 May 1979.

Magma intrusions

Some types of volcanic activity never actually reach the earth's surface. Instead, the magma is forced under pressure into cracks in the surrounding rock, where it cools down and solidifies. This is known as **intrusive volcanicity**.

The largest intrusive feature is the **batholith** – a huge dome-shaped mass of granite, that can be up to 1000 kilometres across. After millions of years, erosion can strip away the overlying rock, to expose the batholith beneath. This has happened in Devon and Cornwall, where the exposed parts of the batholith form bleak granite moors, such as Dartmoor and Bodmin Moor. You can see these in your atlas.

Before the batholith finally solidifies, a number of smaller features are usually intruded into the surrounding rock. These are looked at in Exercise 2.4.

Exercise 2.4

1   On a copy of Figure 2.10, label the features lettered A-F with the following captions, using Table 2.2 to help you.

laccolith
sill
dyke forming a trench at the surface
part of an underlying batholith
sill forming an escarpment at the surface
dyke forming a ridge at the surface

laccolith	a small dome-shaped blister of solidified magma, forcing up the overlying rock layers.
sill	solidified magma which has been injected in sheets along the bedding planes of a sedimentary rock. Sills at an angle are often exposed by erosion to form a steep escarpment. The Great Whin Sill, Northumberland, is an example, and was used by the Romans as a natural defence line during the building of Hadrian's wall.
dyke	solidified magma which has been injected through faults in the rock. They sometimes occur in great numbers, when they are called a dyke swarm. After dykes have been exposed by erosion, they can form trenches if the dyke is more easily eroded than the surrounding rock, or ridges if the dyke is not so easily eroded as the surrounding rock. An example is the Cleveland Dyke on the North Yorkshire Moors, quarried extensively for road building.

◀ Table 2.2

▶ Hydrothermal features such as this one in Iceland are great tourist attractions

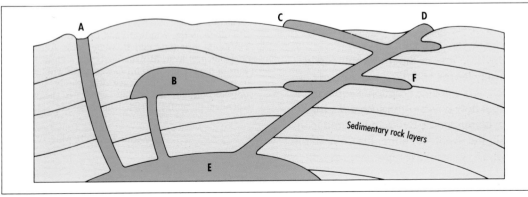

◀ Figure 2.10 Features of intrusive volcanicity

Hydrothermal activity

Hydrothermal activity is found in volcanic areas. Different types of hydrothermal activity can be seen in Figure 2.11. Hydrothermal activity is caused by steam escaping from groundwater which has been heated up by magma rising close to the surface.

Hydrothermal steam can be tapped by boreholes, and led by pipes around towns and villages. The whole of Reykjavik, the capital city of Iceland, is centrally heated in this way. The steam can also be used to drive turbines to generate electricity. Many power stations in Iceland, New Zealand, Japan, the United States and the Soviet Union are powered by hydrothermal energy.

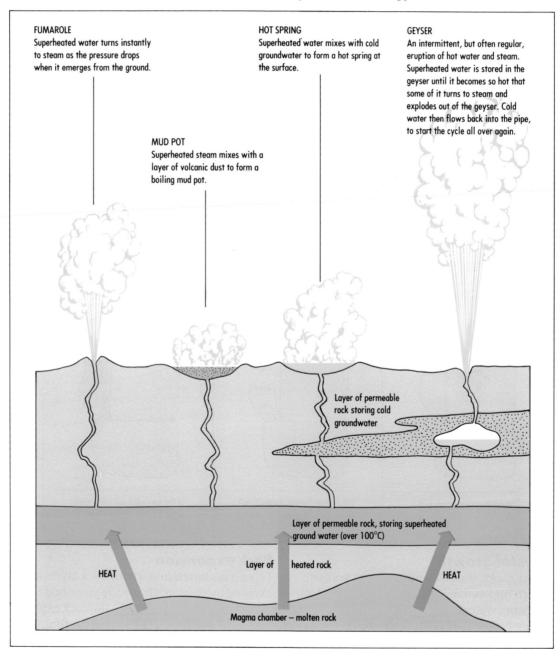

FUMAROLE
Superheated water turns instantly to steam as the pressure drops when it emerges from the ground.

HOT SPRING
Superheated water mixes with cold groundwater to form a hot spring at the surface.

GEYSER
An intermittent, but often regular, eruption of hot water and steam. Superheated water is stored in the geyser until it becomes so hot that some of it turns to steam and explodes out of the geyser. Cold water then flows back into the pipe, to start the cycle all over again.

MUD POT
Superheated steam mixes with a layer of volcanic dust to form a boiling mud pot.

Layer of permeable rock storing cold groundwater

Layer of permeable rock, storing superheated ground water (over 100°C)

Layer of heated rock

HEAT

HEAT

Magma chamber – molten rock

▶ Figure 2.11
Hydrothermal activity caused by superheated water in the ground

Exercise 2.5

1. Use Figure 2.11 to name the hydrothermal feature shown in the photograph.
2. Look back at Figure 1.7. What is it about the geology of Iceland, New Zealand, Japan and parts of the United States that enables these countries to use hydrothermal energy for electric power generation?

THREE

Weathering

Rocks are formed in the earth under tremendously high pressures and temperatures. As the earth's surface is eroded (worn away) by the action of water and ice, the rocks underneath are uncovered. They are exposed to air, water and constantly changing temperatures. All these attack the rocks, causing them to rot away. This process is known as **weathering**. Weathering can be classified into two types – **physical weathering** and **chemical weathering**.

Weathering processes

Physical weathering happens when rocks disintegrate (break up) because they are under stress.

Frost action

When water freezes it expands. If the water is in a rock cavity, this expansion forces the rock apart, as Figure 3.1 shows. This is called **frost shattering**. The bits that are broken off (debris) collect in a fan shape at the bottom of the cliff, to form **scree**.

▲ Spoil heaps produced by the china clay works at St Austell, Cornwall

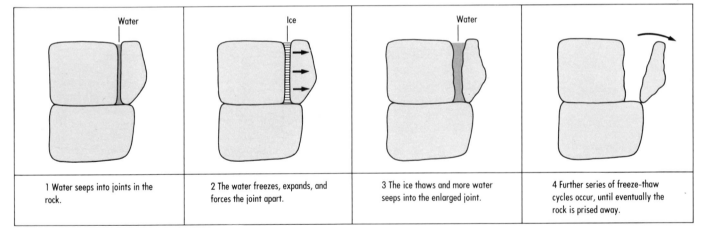

1 Water seeps into joints in the rock.

2 The water freezes, expands, and forces the joint apart.

3 The ice thaws and more water seeps into the enlarged joint.

4 Further series of freeze-thaw cycles occur, until eventually the rock is prised away.

▲ Figure 3.1 Frost shattering

Crystal growth

Many rocks, such as sandstone, are **porous**, which means that they have air spaces which can store water. This water contains mineral salts which have been dissolved out of the rock. As the water evaporates, the salt crystals grow steadily inside the surface layers, pushing off minute flakes of rock. This process is called **granular disintegration**.

In desert areas, the process is so rapid that the whole outer skin of the rock can sometimes peel away, like the skin of an onion. This is called **exfoliation**, and leaves rounded boulders and dome-shaped rocky outcrops.

Rock expansion

This occurs when underlying rock layers are exposed by erosion. Originally squashed under tremendous pressure, the rock expands and cracks in a process called **unloading**.

Chemical weathering occurs when rocks are decomposed (broken down) by a chemical reaction.

Oxidation

Many metals and metallic minerals in rocks combine easily with oxygen to form another substance. Iron is a good example. It reacts quickly to form iron oxide, which causes a red-brown staining on the surface of the rock.

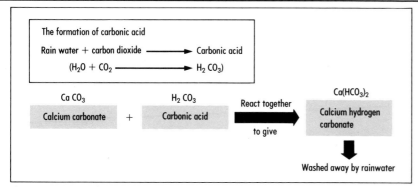

▲ **Figure 3.2** Weathering of carboniferous limestone by carbonation

Carbonation

When rain falls through the air, it absorbs carbon dioxide. By the time it reaches the ground it has become a weak carbonic acid (Figure 3.2). As this slightly acid rainfall soaks into the soil, it absorbs an even greater amount of carbon dioxide from decomposing plant matter. The resulting carbonic acid will quickly attack and dissolve away any rock which contains more than about 50 per cent calcium carbonate, leaving very little behind. This process is called carbonation.

Limestone rocks, which are mainly calcium carbonate, are subjected to particularly rapid rates of carbonation (pages 26 and 27).

Hydrolysis

Other rock minerals are not dissolved by carbonic acid and water, they combine with them. They then break down into other chemical forms. This process is called hydrolysis. A particularly important form of hydrolysis occurs when water combines with granite to form sand and clay which are two very important materials found on the earth's surface.

▼ Figure 3.3

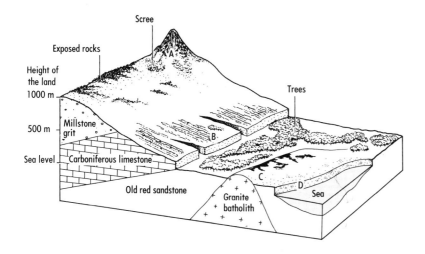

Exercise 3.1

Look at Figure 3.3. This shows a region which has four types of rock, and abundant rainfall. Temperatures vary above and below freezing point every day for four months of the year, wherever the land is over 500 metres above sea level.

1   Link the following captions to the points labelled A-D on Figure 3.3.
    a) An area where hydrolysis is converting the rock to clay and sand.
    b) An area where extensive frost shattering takes place with scree being produced at the cliff bottom.
    c) An area where carbonation is very active, dissolving the rock away.
    d) An area where the rock is liable to salt crystal growth.
2   Write a paragraph giving reasons for your answers.
3   Imagine that the area has been moved by continental drift, and that the whole region has now become a desert. Some moisture, however, is blown in from the sea to form dew in coastal regions in the early morning.
    a) Which weathering processes mentioned in question 1 are likely to stop, and why?
    b) Which weathering process mentioned in question 1 is likely to continue, and why?
    c) Name one weathering process that may now become dominant in the region, and explain how this may change the shape of any exposed rocks.
4   The photograph on this page shows how important weathering is in the production of basic materials such as sand and clay. In this case, hydrolysis of a granite batholith in Devon and Cornwall has produced important deposits of china clay, or **kaolin**. Use your school library to find out about the china clay industry. Write an essay describing the importance of china clay, how it was formed and the history of the industry in Devon and Cornwall. The photograph also shows the devastating effect of this industry on the landscape. In your essay, include an account of the effect of the china clay industry upon the local environment and suggest ways in which it might be controlled.

Acid rain – weathering by pollution

Any rainfall with a pH value of less than 5.6 is considered to be acid rain, (Figure 3.4). Acid rain is caused by the burning of fossil fuels in major industrial areas. Invisible sulphur dioxide and nitrogen oxide fumes rise into the air and are then blown along in the upper atmosphere, sometimes for many thousands of kilometres. During this journey, sunlight causes chemical reactions which convert the gases into sulphuric and nitric acids.

Raindrops and snowflakes absorb these acids, and then fall on to the earth's surface as acid rain. You can see the amount of the acid rain falling on Europe in Figure 3.5. The problem has become much worse over the last twenty years, partly because we use more fossil fuels and partly because we now build taller chimneys. These chimneys reduce local levels of air pollution, but only by sending the fumes much higher into the atmosphere.

Acid rain has several harmful effects. One is that it accelerates weathering. Throughout the world, unique and irreplaceable historic buildings are rapidly being weathered away. These include the Parthenon in Greece, and the Taj Mahal in India. In London, virtually every historic building is affected. For example, carbonation of the limestone of St Paul's Cathedral has increased. In some places, the Cathedral has lost a two centimetre layer of rock. Modern buildings and bridges are not immune from attack either. It is possible that the lifetime of some concrete structures has been halved.

▲ Part of the Black Forest in West Germany, killed by acid rain

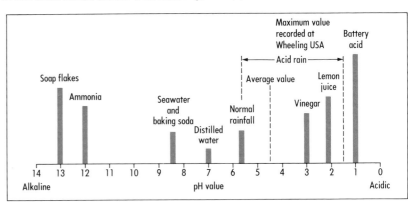

▲ Figure 3.4

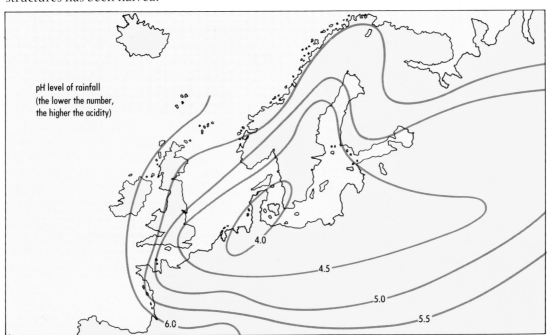

pH level of rainfall
(the lower the number,
the higher the acidity)

◀ **Figure 3.5** Acid rain levels in Europe

Another effect of acid rain is the steady destruction of many soils. Ordinary rain water constantly washes through the soil, removing some minerals and depositing them lower down – a process called **leaching**. Acid rain, however, is more reactive and accelerates the process, causing trees to die (Figure 3.6). Whole forests in Canada, Scandinavia, and Western Europe have been affected in this way. Some people think that in the future, farmland will also be affected.

The worst thing about acid rain is that the countries suffering the worst damage are not necessarily those causing the problem. For example, acid rain caused by the industrial centres of Western Europe falls on Scandinavia. International co-operation is therefore essential to reduce pollution. Sulphur could be removed from coal and oil before it is burned. New burners which use injected limestone to remove the remaining sulphur dioxide could be installed. Filters could take out any remaining gases in the chimney.

Unfortunately, although many governments have now officially recognised that there is a problem, not many are doing anything about it. At present, only Japan has a serious anti-pollution programme, although other countries, such as Britain, have put forward proposals.

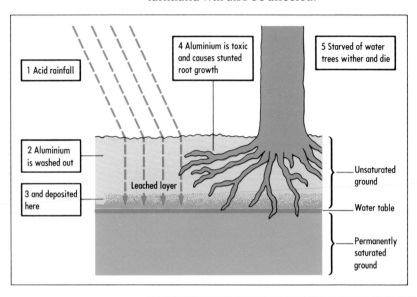

▲ Figure 3.6 The death of a tree due to excessive leaching caused by acid rainfall

1 Acid rainfall

2 Aluminium is washed out

3 and deposited here

Leached layer

4 Aluminium is toxic and causes stunted root growth

5 Starved of water trees wither and die

Unsaturated ground

Water table

Permanently saturated ground

Why the Taj Mahal is looking a little off-colour these days

Growing industrial pollution is slowly discolouring the gleaming Rajasthani marble of the Taj Mahal. Its delicate jewelled inlay is slowly pitting and, in some places, the marble has turned yellow-brown. The red sandstone of some of its adjoining buildings is beginning to flake. Conservationists continue to fear that the entire magnificent structure, built by the grief-stricken emperor Shah Jahan as a tomb for his wife, could be totally discoloured in 50 years.

Meanwhile, on the outskirts of nearby Mathura, the grey chimney stacks of a new state oil refinery continue to belch fumes into the sky. Each hour they emit 25-30 tonnes of sulphur dioxide, and 100 tonnes of nitrogen oxides. The north westerly winds from October to March transport these pollutants direcly from Mathura to Agra, and the Taj Mahal.

The oil refinery is not the only source of pollution. Late last year, the Archaeological Survey of India was responsible for closing down two antiquated coal fired power stations in Agra, and succeeded in having the city's trains switched from coal to diesel. But neither it nor the Indian government has scored any success in closing down or relocating 250 coal burning foundaries, all privately owned. They are a major source of employment for the city – some are as large as factories, others are set up in back yards, where open furnaces create noxious fumes.

Mary Anne Weaver
Adapted from the *Sunday Times*,
17 July 1983

Exercise 3.2

Read the article about the Taj Mahal and then answer the following questions.

1   a) The Taj Mahal is located close to the Indian city of Agra. Use an atlas to write a sentence describing the location of this city.
    b) Name the rock type used to build the Taj Mahal.
    c) Name the rock type used to build the adjoining buildings.
    d) Describe the effects of recent weathering on both these rock types.
2   a) Which weathering process in particular will be speeded up by an increase in rainfall acidity?
    b) Describe the industries which have caused this increased acidity.
    c) Suggest why the Indian government has been reluctant to relocate these industries in order to prevent further damage.
3   Write a paragraph describing some of the short and long term measures which could be introduced to prevent further damage to important buildings.
4   Is there any evidence of buildings being weathered away in your local area? You could carry out a photographic survey and make a wall display, describing and explaining the effects you have observed.

Limestone scenery – a weathered landscape

Weathering is at its most dramatic in areas of carboniferous limestone, where it produces a unique landscape known as **karst scenery**. The rock is **permeable**, which means that water rarely stays on the surface, but can pass through it by way of joints and bedding planes. The result is rapid carbonation (Figure 3.2), which produces the unique landscape features shown on this page.

The map shows a flat upland limestone area, which ends in an abrupt cliff where it meets a band of shale 1 kilometre from the southern edge of the map. The cliff runs virtually straight across the area from west to east, but is broken in the middle by a dramatic steep sided valley, 1.5 kilometres long. A stream emerges from a cave at the northern end of this valley, and runs south off the edge of the map.

In the south west corner of the limestone upland lies a large area of bare rock, approximately one square kilometre in size. On the other side of the steep sided valley, however, there are three funnel shaped depressions in the ground.

A band of clay 1 kilometre wide runs on top of the limestone across the northern edge of the map. Because this rock is impermeable, a stream is able to run from the middle of the northern edge of the map southwards. However, when it reaches the limestone, it suddenly disappears down a deep vertical shaft. This water eventually reappears from the mouth of the cave 2.5 kilometres away.

Another stream appears directly out of the bottom of the limestone cliff 1.5 kilometres from the eastern edge of the map.

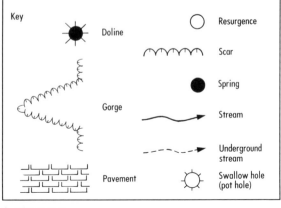

Key

Doline

Gorge

Pavement

Resurgence

Scar

Spring

Stream

Underground stream

Swallow hole (pot hole)

▲ Figure 3.7

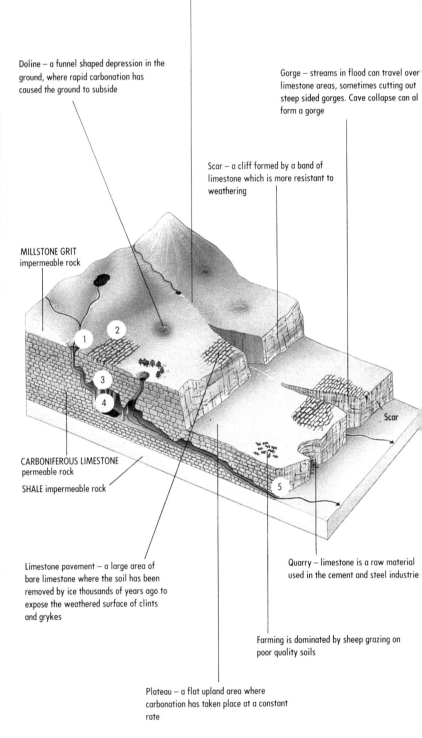

Limestone areas have a thin soil because carbonation does not leave behind any significant soil forming materials such as sand and clay. Vegetation is therefore limited to rough grassland and shrubs

Dry valley – these once contained rivers, before carbonation enlarged the joints sufficiently for water to flow entirely underground, a process which probably took 50 million years

Doline – a funnel shaped depression in the ground, where rapid carbonation has caused the ground to subside

Gorge – streams in flood can travel over limestone areas, sometimes cutting out steep sided gorges. Cave collapse can al form a gorge

Scar – a cliff formed by a band of limestone which is more resistant to weathering

MILLSTONE GRIT impermeable rock

CARBONIFEROUS LIMESTONE permeable rock

SHALE impermeable rock

Scar

Limestone pavement – a large area of bare limestone where the soil has been removed by ice thousands of years ago to expose the weathered surface of clints and grykes

Quarry – limestone is a raw material used in the cement and steel industrie

Farming is dominated by sheep grazing on poor quality soils

Plateau – a flat upland area where carbonation has taken place at a constant rate

▲ Figure 3.8 Limestone scenery – a weathered landscape

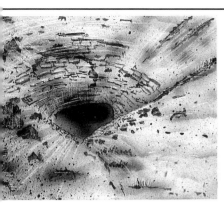

1) Water goes underground at a **swallow hole** (pot hole). This is a joint which has become enlarged by very active carbonation as the water swirls down it, forming a deep vertical shaft.

Pot hole

2) Rain water etches out the joints as it seeps through them. These enlarged joints are called **grykes**. The blocks of limestone in between are called **clints**.

Limestone pavement

3) Once the water has found its way underground, it sinks vertically down to the water table, and then flows horizontally along bedding planes in the rock. The bedding planes are hollowed out to form a system of **caves**. Caves are very old, and owe their size to roof collapse and ordinary stream erosion, once carbonation has enlarged the original passage.

Inside a cave

4) Rainwater seeping down from above drips from the cave roof. Since carbonation has saturated the water with calcium hydrogen carbonate, the drips deposit calcite on both the roof and the floor of the cave as they evaporate. This forms long thin **stalactites** on the roof and short stumpy **stalagmites** on the cave floor. these sometimes join together to form a **pillar**.

Stalactites and stalagmites

Exercise 3.3

In this exercise you are asked to construct a map of a limestone area. You will need to know how to identify all the features shown on this page.

1   a) Draw a 12 cm square in your exercise book.

    b) Using a scale of 2 cm to 1 km, complete your map using the description of the area given in Figure 3.7. Identify each of the limestone features mentioned in the passage and draw them on your map using the symbols shown in the key. Use a pencil and blue colour.

2   a) Label each of the limestone features that you have drawn.

    b) Label the three types of rock shown on your map.

    c) Write a letter P in the centre of the limestone plateau, C/G in an area where you would expect to find clints and grykes, and S/S at a place where you might find stalactites and stalagmites.

    d) Name one limestone feature that you have not marked on your map.

5) The stream finally emerges from the limestone at the joint where it meets impermeable rock. This forms a **spring**. If the stream exits by way of a cave, it is known as a **resurgence**.

Malham Cove

FOUR

Hydrology

Hydrology is the study of water – how it circulates around the earth, and the way in which it is used.

Eighty seven per cent of the world's water is stored in the oceans. This water contains mineral salts which have been weathered out of rocks and brought down to the sea by rivers. These salts make sea-water undrinkable. When heat from the sun evaporates the water, the salt impurities are left behind. The pure water vapour then cools down and condenses out into the clouds. Blown along by the wind, the clouds eventually produce snow and rain. Some of this falls on to the land, where it drains into rivers and is returned to the oceans. This cycle is shown in more detail in Figure 4.2. It is called the **hydrological cycle**.

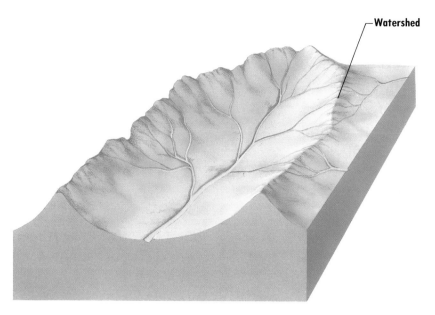

▲ Figure 4.1 A drainage basin

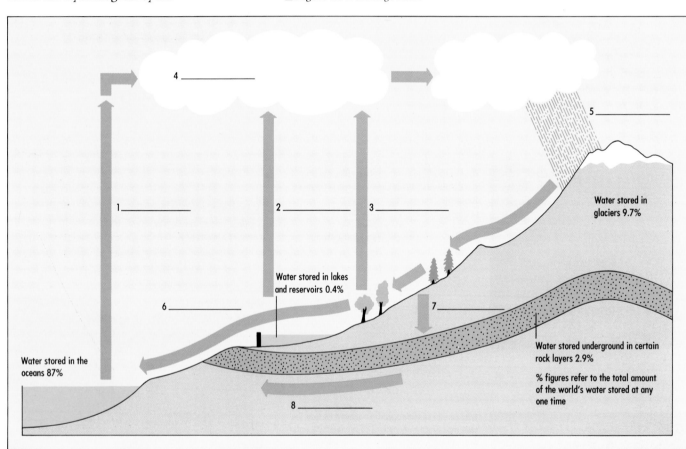

▲ Figure 4.2 The world hydrological cycle

Drainage basins

Rain water drains from the land in a series of eroded bowl-shaped depressions known as **drainage basins** (Figure 4.1). The dividing line between two drainage basins is known as a **watershed**. This normally runs along the crest of the hills which separate the drainage basins. Drainage basins vary enormously in size. Figure 4.3 shows how the drainage basin of the Mississippi in North America covers more than 2.5 million square kilometres of land. Water from one part of this basin can take weeks to travel downstream. Within the Mississippi basin, however, are a number of much smaller ones, such as the Ohio.

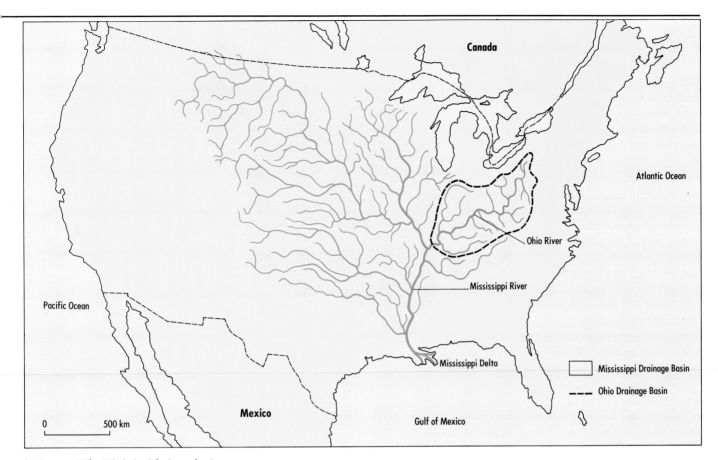

▲ Figure 4.3 The Mississippi drainage basin

▲ River traffic on the mighty Mississippi at New Orleans

Exercise 4.1

1 a) Copy Figure 4.2 and label the arrows with the correct captions from the following list:

Water evaporates from rivers and lakes.

Some water percolates down into underground rock layers.

Water evaporates from the oceans, leaving salt behind.

Rivers return water to the oceans.

Water vapour cools down and condenses out into clouds.

Water transpired by plants.

Underground water is eventually returned to the oceans.

Water falls back to the earth as rain or snow.

b) Colour the arrows red if the movement of water is powered by the sun and blue if the movement of water is powered by gravity.

2 a) Turn to an atlas map of South America. Trace the coastline, and the River Amazon and its tributaries. Mark the likely watershed of the Amazon river system with a dashed line and shade in the drainage basin.

b) Compare the size and land use of the Amazon drainage basin with that of the Mississippi drainage basin.

The drainage basin system

Figure 4.4 shows the movement of water through a typical small drainage basin. When a number of distinct processes are linked in this way, we call the whole thing a system. Water enters the drainage basin as rain or snow, which is called **precipitation** (the input) and leaves via the river channel (the output). However, some of the precipitation never reaches the ground. The leaves of any vegetation growing in the basin catch a great deal of it. This is called **interception**. The leaves hold the water for a short time before it is evaporated back into the atmosphere. This is called the **interception store**.

When the interception store is full, water falls to the ground and begins to soak into the soil. This is called **infiltration**. If the rain is falling very quickly, the soil may not be able to soak up the water fast enough, and water will begin to flow over the surface as **overland flow**. This is in fact quite a rare event except in urban areas, where there is a lot of concrete and tarmac creating a solid ground surface.

As the water infiltrates into the soil, it meets a more compact layer. This makes it flow sideways as **throughflow**. This slow release of water from the soil store keeps the river flowing long after rainfall has finished. Some water from the **soil store** will be taken up by plant roots and **transpired** back into the atmosphere.

When the remaining water reaches the bottom of the soil layer, it begins to move vertically down through the rock. This is called **percolation**. Constant percolation creates a large underground store of water, called the **groundwater store**. The dividing line between this permanently wet rock layer and the dry area above is called the **water table**. Once the percolating water reaches the ground store, it begins to flow sideways through the rock towards the river. This is called **groundwater flow**. This water keeps the river flowing even in times of drought, when no rain falls. However, as the groundwater store is used up, the height of the water table falls. Once it is below the height of the river bed, the river will dry up.

Water therefore reaches the river channel in three different ways – by overland flow, throughflow, and groundwater flow. Once in the river, most of the water is carried out of the drainage basin as **channel flow**. Some may remain in the basin for a long time, however, trapped in lakes or reservoirs as a **surface store**.

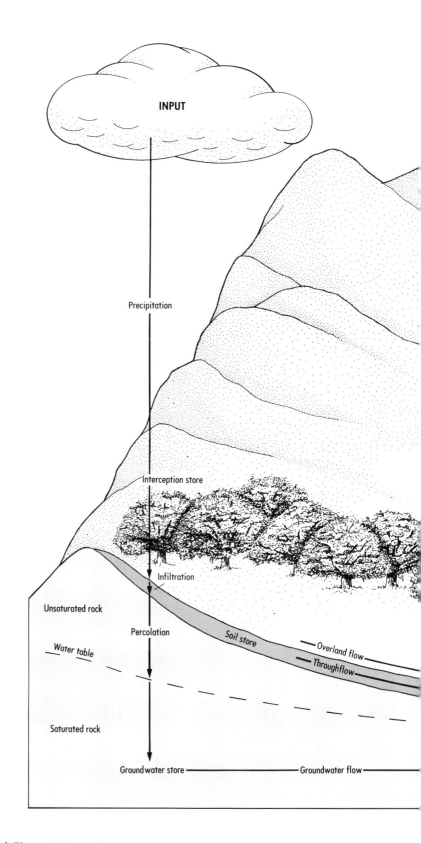

▲ Figure 4.4 The drainage basin system

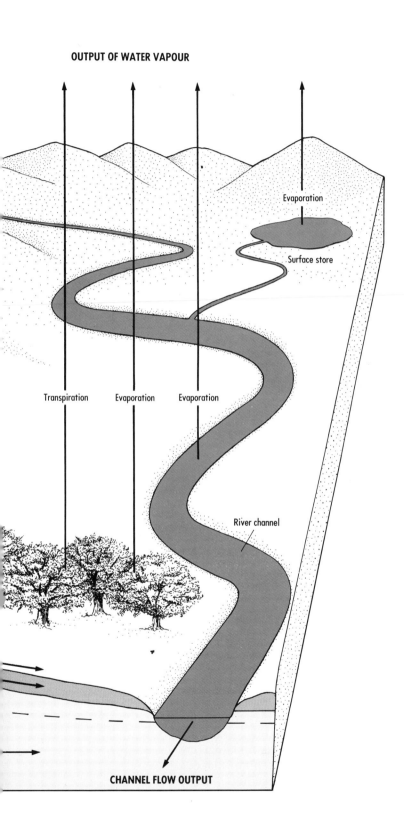

OUTPUT OF WATER VAPOUR

Evaporation

Surface store

Transpiration   Evaporation   Evaporation

River channel

CHANNEL FLOW OUTPUT

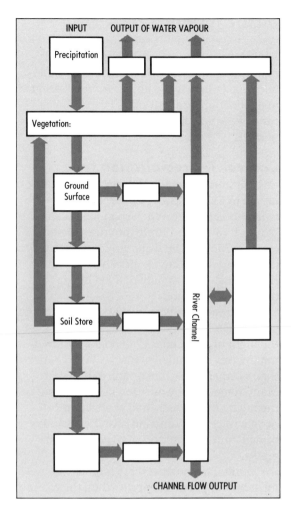

▲ **Figure 4.5** Systems diagram showing water flow through a drainage basin

Exercise 4.2

1   On a copy of Figure 4.5, complete the boxes with the following terms, using Figure 4.4 to help you:

throughflow	groundwater store
overland flow	percolation
groundwater flow	infiltration
interception store	transpiration
evaporation	surface store

2   **a)** Imagine that the area shown in Figure 4.4 has been completely deforested following a boom in the timber industry. Describe what will happen to the outputs of transpiration and evaporation. Give reasons for your answer.

b) Redraw your systems diagram to show the effects of these changes upon river flow.

c) What will be the effect of this change in river flow upon people living near the river downstream?

31

Controlling the drainage basin system

The rivers in some drainage basins rise very rapidly after a storm, often causing very bad flooding. Other rivers, however, never seem to flood. There are four main factors which control the way in which a river reacts to rainfall.

Control 1: precipitation type

Intense storms, or **cloudbursts**, produce a lot of rain in a short space of time, and create the worst flood hazard. When so much rain falls so fast, it cannot infiltrate into the soil rapidly enough. Most of the water flows across the land as overland flow. It quickly enters the river, which rises rapidly to flood height, causing a **flash flood**.

Flash floods can be very destructive. One occurred in Nelson County, West Virginia, on the night of 19 August 1969. Eight hundred millimetres of rain fell in just six hours – more than normally falls on London in one year. Many rivers rose by over ten metres. When the raging floodwaters had gone down 125 people were dead, and farms and villages lay buried under ten metres of river mud and gravel.

Cloudbursts, however, are rare events. Most flooding tends to occur when the ground has become saturated after prolonged rainfall. This prevents infiltration and causes overland flow.

Snowfall can also be a problem. When temperatures rise and the snow melts, the ground beneath often remains frozen, so that the water cannot infiltrate through it. Large volumes of meltwater flow across the surface. After the heavy snowfalls in Britain of January 1982, melting snow flooded the Ouse valley, damaging property in York and inundating valuable farmland.

Control 2: land use type

Vegetation in the drainage basin intercepts rainfall and stores it on the leaves before evaporating it back into the atmosphere. Tropical rain-forest, for example, can intercept as much as 30 per cent of any rainstorm, but crops only capture fifteen per cent because they are planted in rows with bare soil between. Vegetation therefore helps to prevent flooding.

The amount of interception varies with the seasons. In winter, when much vegetation loses its leaves, there is less interception. In arable farming areas, where fields are often left bare in winter, sometimes there is no interception at all.

Urban land use in particular increases the risk of flooding. Water can not infiltrate into concrete and tarmac, and roofs and gutters channel stormwater straight into underground drains. The water flows into the nearest river, quickly raising it to flood height.

Control 3: soil type

Sandy soils have large **pore spaces**, usually greater than 0.2 millimetres in diameter. This allows rapid infiltration; the soil soaks up the rainfall very quickly. Clay soils, however, have much smaller pore spaces, usually under 0.002 millimetres in diameter. Very slow infiltration encourages overland flow, creating a risk of flooding.

Soil type also controls the speed of throughflow. On average this varies between 0.5 and 30 centimetres per hour, and due to its slow speed it rarely causes flooding. However, sometimes throughflow makes minute cracks in the soil, creating small underground channels called **pipes**. Water can travel quickly through these, increasing the risk of flooding. Farmers produce the same effect when they artificially drain the land by laying drainage pipes in the ground, or by digging drainage ditches.

▲ Tree felling can have a major effect on water flow in the drainage basin

Control 4: rock type

Certain rocks allow water to pass through them. These are called **permeable rocks**, and there are two types. **Porous rocks** contain minute pore spaces which can fill with water. Examples of porous rocks are sandstone and chalk. **Pervious rocks** are not porous, but allow water to flow along cracks in the rocks. An example is limestone. Rocks which are neither porous nor pervious are called **impermeable rocks**.

Permeable rocks tend to prevent flooding because stormwater is able to percolate rapidly into a large groundwater store. In fact, surface rivers only run over these rocks when there is an impermeable layer on top, such as clay, or when the water table has reached the surface after a long period of rainfall.

Exercise 4.3

1 a) Look at Figure 4.6. Briefly describe the distribution of soil and rock types in the area.
   b) Describe the change in land use between 1935 and 1985.

2 What effect will each of the following factors have upon water movement in the area?
   rock type
   soil type
   ploughing up pasture for arable land
   artificial drainage of farmland
   removal of woodland for urban and parkland development

3 Suggest some of the measures you would use to slow down the speed of water movement through this drainage basin in 1985.

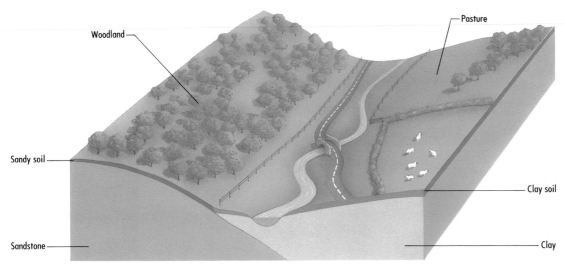

1935

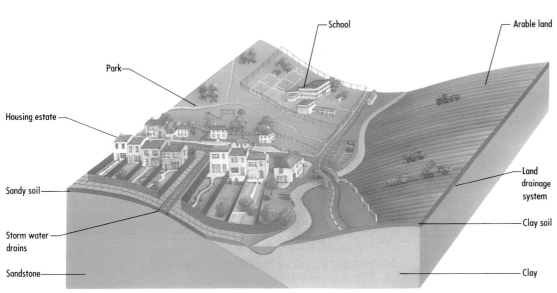

1985

◄ Figure 4.6 The changing land use of a small drainage basin

Flooding and flood control

▲ Buildings collapsing during the Lynmouth flood

One of the most common causes of flooding is an intense rainstorm which simply swamps the drainage basin system with a huge amount of excess water. This was the cause of the Lynmouth flood disaster in 1952. Two weeks of heavy rain had filled up the soil and groundwater stores, and then there was a sudden cloudburst.

Thirty-four people died as houses were swept away, and parts of the village were buried under 200 000 tonnes of mud and rock.

Rivers with a high flood risk can be detected by drawing a **hydrograph** (Figure 4.7). This plots the way the river flow rises after a rainstorm. The river flow is measured as discharge. The discharge of a river is the volume of water passing a point on the river at a given time.

Rivers with a high flood risk have **flash response** hydrographs. In this case, river flow rises rapidly after the rainstorm, creating a high **peak discharge**. The opposite of this is a **lag response**, when river flow is slow to rise. Pages 32 and 33 showed how the type of response is controlled by land use, soil type and rock type in the drainage basin.

There is growing evidence that the risk of flooding is increasing in many areas of the world. In the United States, for example, the death toll from floods has tripled since the 1940s. The main reason for this is a change of land use within many drainage basins. New urban land has been developed and lots of trees have been felled (deforestation).

Rivers with a high flood risk can be controlled in four main ways:

- Build flood banks down either side of the channel to contain the floodwater.
- Straighten, widen and deepen the river channel, so that it can carry more water.
- Line the channel with concrete to reduce friction and increase river speed.
- Build dams to catch the floodwater, and then release it more slowly.

The first three of these solutions are relatively inexpensive and easy to do. However, they can often create new problems elsewhere. They may prevent flooding and erosion in one part of the river, but at the same time they increase the risks of these for people living downstream, because there is now more water and it is flowing faster. The best solution is often good **land management**. Forestry, farming and building in the drainage basin are all carefully controlled so that rapid run-off into the river is prevented.

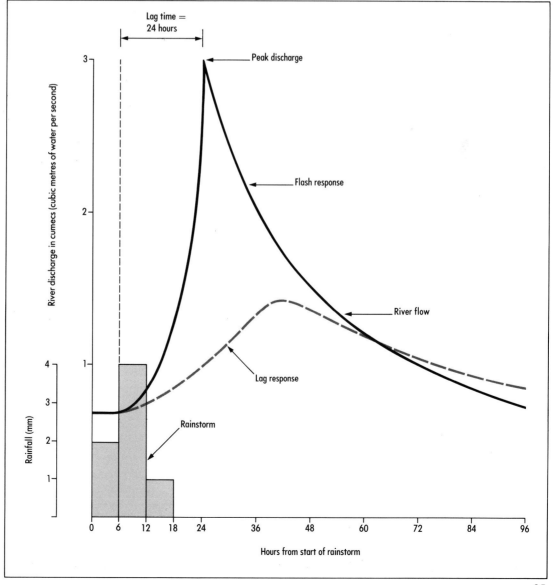

► **Figure 4.7** A typical hydrograph, showing the way in which a river rises after a rainstorm

Case study:
The Brent flood,
16 August 1977

Flooding in London makes most people think of the River Thames and its flood barrier. However, a much greater risk comes from the five smaller tributary rivers of the Thames which flow through the city. One of these is the River Brent.

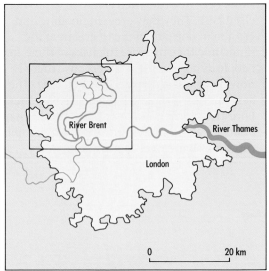

▲ Figure 4.8

The flood risk

The River Brent drains water from 173 square kilometres of north-west London (Figure 4.8). About 50 per cent of its drainage basin is covered by residential land use, and a further 20 per cent is covered by shops and offices. The remaining land is used for parks, industry, and important road and rail links. Much of this development occurred very rapidly in the 1930s, and little thought was given to the risk of flooding. Property was built on low-lying floodplains close to the river banks. Storm drains now channel water directly into the river system. Paved surfaces prevent rain water from reaching the natural soil and groundwater stores in the drainage basin. Rubbish dumped into the river channel makes the problem worse. It is picked up by the water, and then jammed up against low lying bridges which slows down the water flow. There is also growing evidence of a change in the local climate, with an increase in the number of intense thundery rainstorms over the last 40 years.

▲ Brent in flood, August 1977

▲ Brent today

The risk of flooding is therefore growing and there is a risk of a serious flood about once every 40-50 years. A study in 1973 suggested that a flood prevention scheme for the River Brent would cost £17 million, and take twenty years to build.

The flood of 16 August 1977

The flood was caused by heavy thundery rain, which sent some rain gauges off the scale. Eleven hundred houses, twenty factories, and many shops and offices were seriously flooded, along with important trunk roads, railway routes and underground stations. Exercise 4.4 investigates this flood in detail.

Exercise 4.4

1 a) Figure 4.9 shows the 60 millimetre **isohyet** for north west London on 16 August 1977. An isohyet is a 'contour'

line joining up points of equal rainfall.
On a copy of Figure 4.9, draw four more
isohyets from 70 millimetres to 100
millimetres.

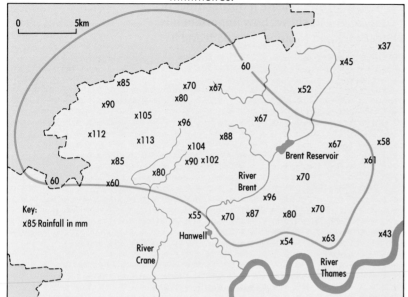

▲ Figure 4.9 An isohyet
map showing rainfall in
north west London on 16
August 1977

b) Briefly describe the pattern of rainfall.

2 a) Using the axes shown in Figure 4.10
complete a hydrograph for the River
Brent using the information given in
Table 4.1.

b) How long did the storm last, and how
much rain fell?
c) Would you describe your hydrograph
as a flash response or a lag response?
Give reasons for your answer.
d) Explain why the type of land use
found in the River Brent drainage basin
is likely to be the cause of this type of
river response.

3 a) Read the article about flooding in the
south. Summarise the factors which
indicate the severity of the flood.
b) Write a short reasoned account of the
flood prevention measures you would
use in a densely urbanised area such as
the River Brent drainage basin.

4 Think about land use in your local area.
Write an illustrated report for your Water
Authority about sites of potential flood
hazards.

Floods disrupt travel in the South

Travel disruption. The storms disrupted road and rail travel and telephone services in southern England and the Midlands (the Press Association reports). Almost 24 hours later, floodwater was still causing difficulties in north and north-west London. Some roads in Greenford were still under 6ft of water, and diversions were set up where the Grand Union Canal overflowed on to the North Circular Road. Some cars on the road were submerged.

Many rivers, including the Thames, were still high last night, and there were fears that further rain would add to the disruption.

The police evacuated more than 30 people from homes in the Greenford area.

Among at least 20 main London roads badly affected by the flooding were Chelsea Embankment, Brent Cross, and Hanger Lane at Ealing.

Train services to and from Euston were subject to delays. Local services from Bedford, St. Albans and Luton, which normally run to Moorgate, were diverted to St. Pancras, and other services north of London were disrupted.

Many Underground stations were out of action and commuters delayed.

Nearly half an inch of rain fell at Heathrow airport last night. At Hayes, several families were evacuated in boats from their homes. They were sent to a school for the night.

The London fire service said it received so many calls for help that it had started to lose count. In the Acton area floods were up to 5ft deep. Abandoned vehicles added to the chaos.

About 70,000 telephones in London were put out of order by flooding. A restricted service operated in north-west and west London and a serious cable fault affected calls within central London.

The London Weather Centre said it was the wettest 12-hour period since August 1971.

Hugh Clayton
The Times Thursday 18 August 1977

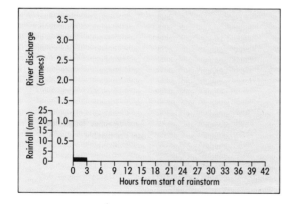

► Figure 4.10

Time from start of rainstorm (hours)	Average rainfall at Hanwell (in mm for each 3 hour period)	Discharge at Hanwell (cumecs)
0	0	0.5
3	2	0.5
6	19	0.7
9	25	1.8
12	23	2.5
15	0	3.0
18	0	3.5
21	0	3.0
24	0	2.5
27	0	1.5
30	0	1.0
33	0	0.8
36	0	0.7
39	0	0.5
42	0	0.5

► Table 4.1

Water supply

Water is vital to everyday life. We not only drink it and use it for many domestic purposes, but it is also required for agriculture and a great number of manufacturing processes. We therefore have to store it so that there is a constant supply.

A common method of storing water is in a **reservoir**. A dam is built to store water during times of surplus and release it during times of deficit. Figure 4.11 shows how important it is to choose the right location for a dam and its reservoir.

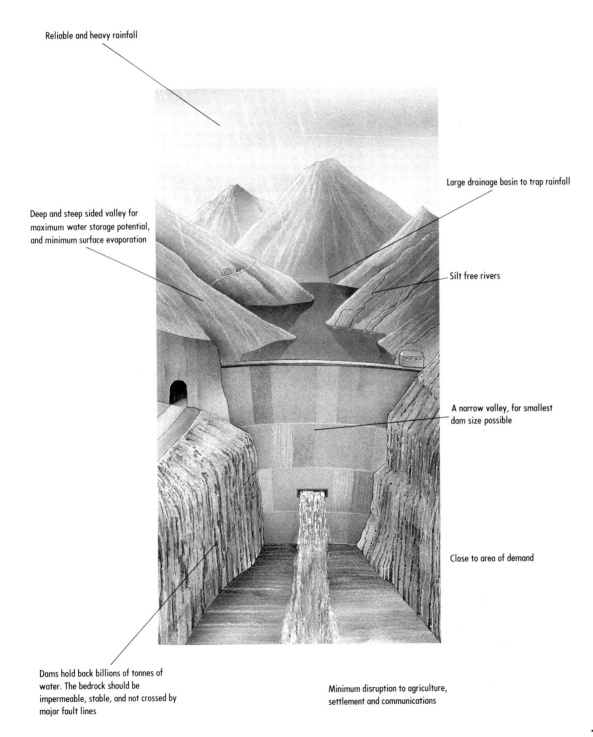

Reliable and heavy rainfall

Large drainage basin to trap rainfall

Deep and steep sided valley for maximum water storage potential, and minimum surface evaporation

Silt free rivers

A narrow valley, for smallest dam size possible

Close to area of demand

Dams hold back billions of tonnes of water. The bedrock should be impermeable, stable, and not crossed by major fault lines

Minimum disruption to agriculture, settlement and communications

◀ Figure 4.11
Requirements for a major dam and reservoir

► A London reservoir during the 1976 drought

Aquifers provide another method of water storage. An aquifer is a huge layer of porous rock, such as sandstone or chalk, creating a large groundwater store. Figure 4.12 shows how the aquifer below London is filled by rain falling on the Chiltern Hills and North Downs. If a well is dug below the level of the water table, water flowing into it will be forced out on to the surface under pressure. This is called an **artesian well**. One artesian well was used to provide the water for the fountains in Trafalgar Square. Since 1850, however, the growing demand for water in the capital has steadily lowered the water table, and the water now has to be pumped out by electricity.

In the future, there will probably not be enough storage to supply all the water we need. Many people wish to avoid building yet more reservoirs in attractive upland areas. An alternative might be to build **barrages**. These are earth dams across the mouths of major river estuaries. Fresh water would collect behind the barrage, creating a huge reservoir. However, the effects of this upon shipping, and the destruction of a natural salt water environment, are both still major problems.

A national water grid, which can transfer water from areas of surplus to areas of deficit (not enough) during a drought, might well be a more useful solution.

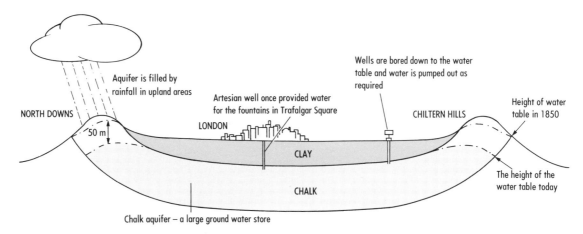

► **Figure 4.12** The London aquifer

Case study:
The Kano River Irrigation Project – a multi-purpose river scheme in northern Nigeria

In an attempt to reduce food imports and increase home production the Nigerian government is encouraging the development of agriculture.

Northern Nigeria has a savanna climate. During the summer, the area receives heavy rainfall, but it can be unreliable. During the winter there is drought. So a large supply of water is needed for irrigation (watering the land). There are plans for getting this supply from many drainage basins, including that of the River Kano. Reservoirs are being built on impermeable granite rock (Figure 4.13).

By far the largest reservoir is that created by the Tiga Dam, which provides irrigation water for 5000 square kilometres of land in the Kano valley. This has changed farming dramatically. There are up to two harvests per year of wheat, maize, cotton and market garden produce. None of these were grown before. The reservoir also stores water for domestic and industrial use in Kano which is an important developing city in northern Nigeria. A 20 000 kWh power station has also been built at the dam site, and fish farming is being encouraged. The dam will also provide flood control on the River Kano during the wet season.

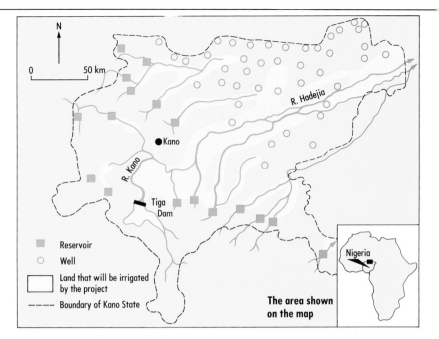

▲ Figure 4.13 The Kano Irrigation Project

But the reservoir has also caused some serious problems. Sixty thousand people had to be rehoused in new villages, and given money as compensation. The Kano river used to flood every year, leaving a layer of silt on the floodplain. This was a natural and free fertiliser. But the dam has stopped this flooding. Farmers now have to buy in expensive foreign fertilisers. The silt is building up in the reservoir, which will create further problems for the future. The reservoir has also brought a major health hazard to the people living close by. They are in danger of catching a water-borne disease called bilharzia, which is common in stagnant water. There is no effective cure.

◄ A new irrigation canal in Kano, Nigeria

In the north and east of the Kano region, water development is quite different. Alternate layers of clay and permeable sandstone make it impossible to construct reservoirs but provide an ideal area for aquifer storage. Wells have been bored into the sandstone, and can provide up to eighteen litres of water per second. This is enough water for domestic village use, but not for irrigation. These wells do cause some problems. Cattle farmers, whose herds used to wander over the area, now want to stay close to the wells. This means the land gets overgrazed, which in turn leads to soil erosion. One solution would be to build more wells, so that the farmers can move their herds about more. But this would be a very expensive solution.

Month	Monthly		
	rainfall (mm)	air temperature (°C)	evapotransp- iration (mm)
January	0	21.5	155
February	0	24.0	140
March	3	27.0	186
April	10	31.0	210
May	63	30.0	217
June	110	29.0	180
July	200	26.5	155
August	310	25.0	155
September	125	26.0	150
October	13	27.5	186
November	0	25.0	150
December	0	22.5	155

▶ **Table 4.2** Climatic data for Kano State, Nigeria.

Exercise 4.5

1 a) Use the information in Table 4.2 to plot the monthly rainfall figures as a series of bars on a copy of Figure 4.14.

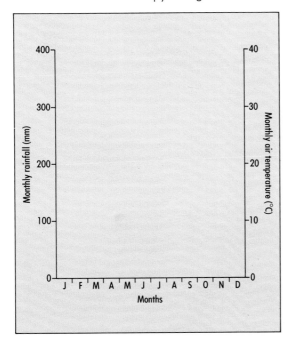

▶ **Figure 4.14** A climate graph for Kano, northern Nigeria

b) Then plot the monthly air temperature figures as a series of points, joining them up with straight lines.

c) Briefly describe the way in which rainfall and temperature vary throughout the year. What is this type of climate called?

d) Find the climate graph of your own area in an atlas. Describe three differences between your own climate and that of Kano.

2 a) Copy Figure 4.15 onto tracing paper. Plot monthly evapotranspiration as a series of red bars. This is the maximum possible amount of water that can be evaporated or transpired in that month.

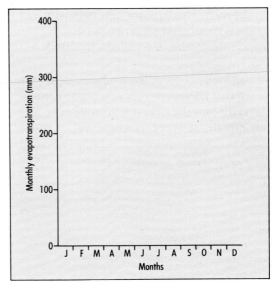

▲ **Figure 4.15**

b) Lay your tracing paper over your completed climate graph. During which months of the year will farmers in Kano need irrigation water, and why?

3 Imagine that you are a member of the FAO (Food and Agriculture Organisation). You have been asked to produce a short illustrated publicity leaflet telling the story of the Kano Irrigation Project. You have space for six photographs. Describe the pictures that you would use, and write suitable captions (about one paragraph in length). You have been told that your leaflet should explain why the project is important for the Nigerian economy, describe how the scheme works, and how it has affected people's lives. But you should also point out some of the problems that have occurred with their likely future solutions.

41

FIVE

Rivers and river valleys

The river valley system

Look at the photograph. It shows a river running through a small upland valley. The rock on the valley side is being broken down by weathering. If the weathered material becomes unstable, gravity and rain will make it fall down into the river, which then carries it away. These events can be represented by a systems diagram, as shown in Figure 5.1.

As the weathered debris is transported by the river, it erodes the river bed. This makes the slope of the valley steeper. So yet more weathered material slips down the valley sides into the river. This exposes new rock which in its turn is weathered away. The systems diagram shows this. It is called a **feedback loop.**

If you look closely at the feedback loop, you will see that it suggests that as more and more debris falls into the river it will erode ever more rapidly, in a process which will accelerate out of control. This of course does not happen. The system is controlled by how much debris the river can actually carry. Too much debris, and the river slows down, preventing further erosion from taking place. Too little debris, and the river flows more rapidly, increasing the rate of erosion. In this way every river reaches a balance between the rate of valley side erosion and the amount of debris transported by the river.

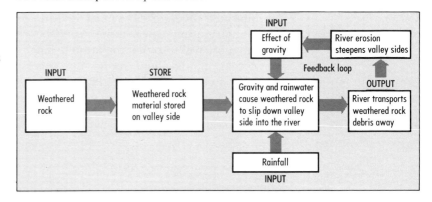

▲ Figure 5.1 A systems diagram for a river valley

◀ The River Aled, North Wales

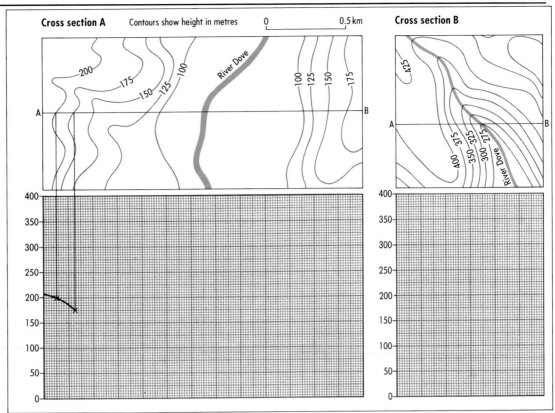

Cross section A Contours show height in metres 0 0.5 km **Cross section B**

◀ Figure 5.2

Exercise 5.1

Figure 5.2 shows two cross sections of the River Dove, in Derbyshire.

1 a) Trace Figure 5.2 and complete cross section A by drawing a line vertically down from each point where the line AB crosses a contour, to the correct height for that contour on the graph. The first two points have been completed for you.
   b) Join all the points with a straight line, and arrow and label the position of the River Dove.
   c) Complete cross section B in the same way.

d) Label one cross section 'A typical flat bottomed lowland valley', and the other 'A typical V-shaped upland valley'.

2 Describe the differences between the two valley shapes under the following headings:
   Shape of valley floor
   Width of valley floor
   Steepness of valley sides
   Width of river channel

3 These differences between upland and lowland valley shapes can be explained by using a systems diagram.
   a) Make two copies of Figure 5.3.
   b) Using Figure 5.1 to help you, complete one diagram for an upland valley, and one diagram for a lowland valley. Fill in the boxes from the captions listed in Table 5.1.

▼ **Figure 5.3** A systems diagram for a _____ valley

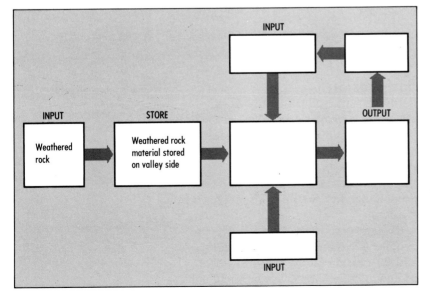

Upland valley system	Lowland valley system
high rainfall	little river erosion
rapid river erosion	slow transport of weathered material
steepened valley sides increase the effect of gravity	low rainfall
weathered material slips rapidly down valley sides	weathered material tends to remain on valley sides
rapid transport of weathered material	gentle valley sides reduce the effect of gravity

▲ **Table 5.1**

c) Use the systems diagrams to write a paragraph explaining why upland and lowland valleys have different shapes.

43

The effect of rainfall on slopes

A heavy rainstorm can wash hundreds of tonnes of soil down any slope, and into the nearest river or stream. It does this in two ways.

Rainsplash

Rainsplash occurs when rainfall hits bare soil. On flat land, this merely shuffles the soil particles around. On a slope, however, soil particles will be splashed further downslope than upslope, as shown in Figure 5.4.

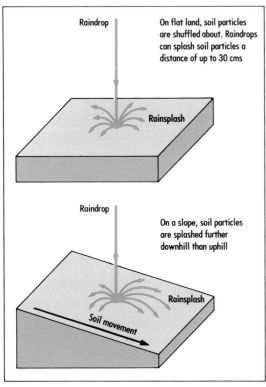

▲ Figure 5.4 Rainsplash

Soil wash

Soil wash occurs when overland flow carries soil particles rapidly down a bare slope which has no vegetation growing on it. At the top of the slope, rain water tends to flow over the land as a continuous sheet. Further down the slope, however, the water begins to run in a series of separate small channels, known as **rills**. The rills concentrate the energy of flowing water, allowing faster erosion. The biggest rills erode the surrounding soil to form **gullies**. Erosion inside the gullies can be fierce enough to remove all the soil right down to the bedrock. At the bottom of the slope, where the gradient is less steep, the water slows down and deposits the debris it is carrying.

▲ Catastrophic soil erosion on newly ploughed land, Rondônia, Brazil

Soil erosion

Rainsplash and soil wash only work effectively on bare slopes. These are mainly found in areas of desert or high mountains. Most slopes have vegetation growing on them.

Leaves intercept the rainfall, and roots bind the soil together. So the soil is protected from erosion.

In many parts of the world, however, people have removed this vegetation cover, and soil erosion is now a serious global problem. In the intensively farmed English Midlands, for example, farmers are ploughing

1 Using the information shown in Table 5.2, plot a scattergraph showing the relationship between rainfall and soil erosion in four parts of West Africa. Figure 5.5 gives you suitable axes and a key.

Place	Average annual rainfall (mm)	Soil erosion (tonnes/hectare/year)		
		forest land	crop land	bare soil
Abidjan (Ivory Coast)	2100	0	45	139
Bouake (Ivory Coast)	1200	0.1	12.5	24
Ouagadougou (Burkina Faso)	850	0.1	0.7	15
Sefa (Senegal)	1300	0.25	7.3	21.3

▲ **Table 5.2** The amount of soil erosion in parts of West Africa

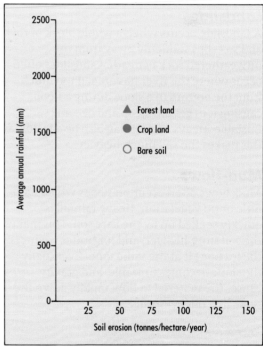

▲ Figure 5.5

up steep grass-covered slopes for arable cultivation. Recent research shows the rate of soil loss has increased from 2.4 tonnes to a staggering seventeen tonnes per hectare per year. In the United States, soil is being removed from some cultivated farmland eight times faster than it is being formed. In the tropical and semi-desert regions of the world, deforestation, overgrazing, and intense rainstorms are the cause of some of the worst cases of soil erosion. (An example of soil erosion in the Himalayas is examined on pages 48 and 49.)

2 a) Describe the relationship between soil erosion and annual rainfall.
   b) Explain how rainsplash and soil wash cause this relationship.
3 a) Describe the relationship between soil erosion and land use.
   b) Explain why some slopes do not suffer from excessive soil erosion.
4 Why do you think soil erosion is increasing in many parts of West Africa? What effect will this have upon the lives of people living in rural areas?

The effect of gravity on slopes

Gravity is constantly tugging away at the soil layer on a slope, putting it under tremendous stress. The soil resists this stress. It gets this strength for resistance from two factors:

- **Friction** between particles of rock and soil. The locking together of individual rock and soil particles like the pieces of a jigsaw puzzle.
- **Cohesion** – the small amount of water present in soil acts like a 'glue', sticking soil particles together.

On a stable slope, the stress of gravity is matched by the strength of the material, and the slope stays at its natural angle of rest. The slope can suddenly give way, however, if erosion makes the angle steeper or if the soil layer is made heavier.

Landslides

A landslide occurs when a whole block of rock and soil slips suddenly away from the unweathered rock beneath. Landslides often occur where the river has eroded too much from the bottom of a slope during a flood, causing it to get too steep and become unstable. Heavy rainfall can also be a factor, as this makes the soil much heavier.

Mud-flows

Mud-flows often occur on slopes which have little or no vegetation. Heavy rainfall is quickly soaked up by the bare soil. This makes the soil store heavier, and lubricates (reduces the friction) it at the same time. Eventually the slope becomes unstable, and gravity causes the material to flow rapidly down the slope like thick porridge.

Soil creep

The slow and undramatic process of soil creep is the most common way in which material moves down a slope. Individual soil and rock particles heat up during the daytime, and expand. When they cool down at night, they contract.

The downhill shuffle

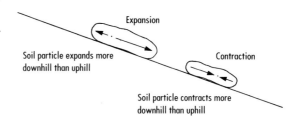

The downhill bounce

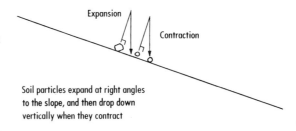

◀ **Figure 5.6** Soil creep

This daily expansion and contraction moves the particles down the slope at an average speed of about one millimetre per year (Figure 5.6). The wetting and drying of soil particles has the same effect. Despite this incredibly slow pace, the whole surface layer of the slope is on the move, and over a long period of time very large amounts of material can change position. You can see evidence for soil creep in Figure 5.7. The constant trampling of cattle and sheep along the side of hill can have the

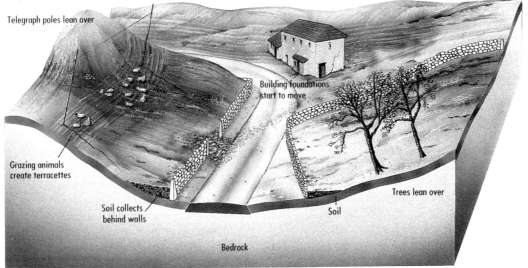

◀ **Figure 5.7** The effects of soil creep

▲ New blocks of flats, on Hong Kong Island

same effect. Rows of small horizontal steps form around the hill as the soil is pushed down. These are called **terracettes**.

Engineers have to take the natural stable slope angle of any material into account when they are both constructing artificial slopes such as cuttings or embankments, and when they are adding weight by building roads or houses. The problems encountered during the building of the Sevenoaks by-pass in Kent provide a good example of what can happen. The road was built across a series of old stablised Ice Age mud-flows. When a cutting for the new road sliced off the bottom of an old mud-flow, this increased the slope angle. The mud-flow started again and began to advance across the road (Figure 5.8). Where an embankment was built on top of an old mud-flow, the increased weight had the same effect, and the reactivated mud-flow began to carry the road away. The by-pass had to be re-aligned at great expense.

By increasing the effect of gravity:

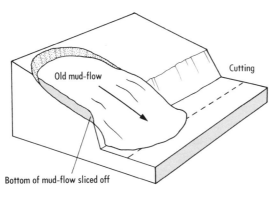

By increasing weight:

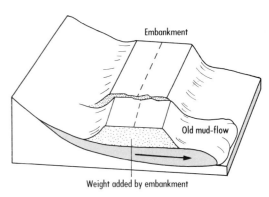

▶ Figure 5.8 How the Sevenoaks by-pass reactivated ancient Ice Age mud-flows

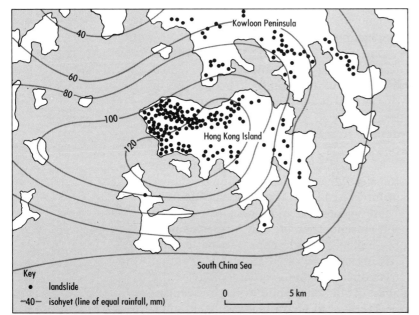

▲ Figure 5.9 Distribution of landslides on Hong Kong Island

Exercise 5.3

Hong Kong is a small island off the South China coast. In a few days in June 1966, a number of major landslides occurred, killing 64 people, and making 2500 people homeless. Similar landslides have occurred frequently since that date.

1   a) Look at Figure 5.9. Is there a relationship between high levels of rainfall and large numbers of landslides?

b) Explain why heavy rain can cause a landslide.

2   a) Look carefully at the photograph. There is a great shortage of suitable building land on Hong Kong island. What evidence can you see in the photograph to support this statement?

b) Explain why this has increased the danger of landslides.

Case study:
Soil erosion in Nepal

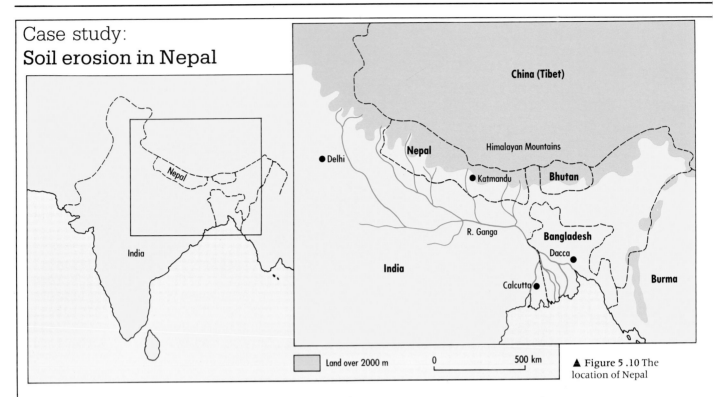

▲ Figure 5.10 The location of Nepal

Nepal is a country about the size of England, situated on the southern slopes of the Himalayan Mountains (Figure 5.10). It has a population of fifteen million, most of whom are subsistence farmers. Families support themselves by farming land scattered in tiny terraced plots around the steep valley sides.

In the past, the amount of land and the size of the population were in balance. However, over the past 30 years, the population of Nepal has doubled. Mounting population pressure has caused severe soil erosion, and soil is now being lost at the rate of twelve tonnes per hectare per year – double the normal rate for the region.

There are four related causes of this severe soil erosion.

In their search for more land, farmers have moved higher up the valley sides. They have begun to terrace land which is really too steep to be engineered in this way. When the heavy monsoon rains fall, the terraces often collapse.

The number of cattle in the area has increased. These provide power for ploughing and transport, manure for fertiliser, milk, and cash from the sale of calves. In an effort to grow more food, the farmers are now using more cattle. The land has become over grazed and severely trampled. This has lowered the infiltration rate of the soil, and increased the rate of overland flow.

The growing demand for firewood and cattle fodder has led to deforestation. Nepal has lost one third of its forests in just ten years. This has lowered the interception rate of the area, causing increased overland flow.

The rain of the monsoon season, which lasts from June to September is very heavy. During these four months, more rain can fall on Nepal than falls on the UK in five years.

Nepal's soil erosion has an effect on neighbouring India. The rivers flowing out of the Himalayas feed the fertile floodplain of the River Ganga in North East India, where ten per cent of the world's population live.

Due to deforestation, the amount of water flowing in the rivers has increased dramatically. Thousands of people die in the annual floods. Productive farmland is lost due to accelerated river erosion, and difficult river crossings disrupt communications.

In order to solve the problem, the Nepalese government has set up a number of Integrated Land Management Agencies. These agencies identify areas of excessive soil erosion. They prevent further damage by building check dams across gullies, fencing off overgrazed areas, and planting trees on bare slopes in order to increase the amount of interception. Such agencies have produced spectacular results, in some

▲ Terracing on steep valley sides in Nepal

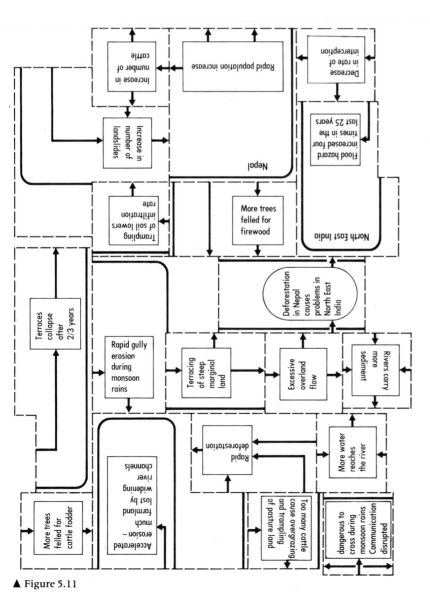

▲ Figure 5.11

cases cutting down the rate of soil loss by ten per cent in just one year.

The Nepalese government has also tried to establish a number of self-help forestry schemes. Villagers are encouraged to set up their own tree nurseries, and plant saplings on bare slopes. They will be able to use the wood for fuel supplies and construction materials. They may also be able to sell some of it.

It may be more difficult to work out long-term solutions. Instead of cutting down trees for firewood, people could use methane gas as a fuel. This is given off by fermenting farmyard manure, and then stored in tanks for domestic use. The slurry from this process is also a good fertiliser. However, the fermentation process requires steady warmth to operate efficiently, so only farmers in the warmer lowlands could use this method. Another answer might be water power, since upland Nepal has many tiny fast-flowing streams. These could power small water wheels and hydro-eleric power stations.

The ultimate solution is of course population control. However, because the people are poor, most of them want large families. They see extra hands as being the only way that they will ever grow enough food.

Exercise 5.4

1 Read pages 48 to 49 before doing this exercise.

2 a) Working in pairs trace a copy of Figure 5.11.

b) Cut out all the pieces along the dotted lines, and reassemble them into a diagram showing how rapid population increase in Nepal has caused increased soil erosion and river flooding. Start by placing the piece labelled 'Nepal' in the top left hand corner. The completed diagram should be a square and a rectangle, linked together with all the arrows matching up. Stick the finished diagram into your exercise book.

c) Briefly describe and explain the sequence of events shown on your diagram.

3 a) List the short-term and long-term solutions to the problem under the appropriate headings.

b) Choose one of these solutions, and explain in detail how it will work.

The power of rivers

Running water possesses a huge amount of energy. Surprisingly, you would find it difficult to stand up in water that was only one metre deep, and running at just three kilometres per hour!

The river has this tremendous energy simply because it flows down from high land to sea level.

The amount of energy in a river is controlled by how much water is actually in the river (its **mass**), and the speed at which it is flowing (its **velocity**). Of these two factors, velocity is the most important, since a small increase in velocity leads to a tremendous increase in energy. Surprisingly, 95 per cent of this energy is used up by the river simply to overcome the friction between itself and the river bed and banks. The remaining five per cent is then used to erode the river channel, and carry the debris away.

Exercise 5.5

1   Look at Figure 5.12 which shows a cross section through a river whose bed and banks have been reinforced with concrete.
   a) What is the maximum velocity of water in the river? What is the minimum velocity?
   b) Describe where the maximum and minimum water velocities are found, and say why you think this pattern occurs.

2   Look at Figure 5.13 which shows a cross section through a river with natural bed and banks.
   a) On a copy of Figure 5.13, draw four isovels at 5 cm/second intervals, starting at 15 cm/second and ending at 30 cm/second. The 35 cm/second isovel has already been drawn in for you.
   b) What is the maximum and minimum velocity of water in the river?
   c) Suggest a reason for any difference you notice between river velocities in the two channels.

Concrete channel

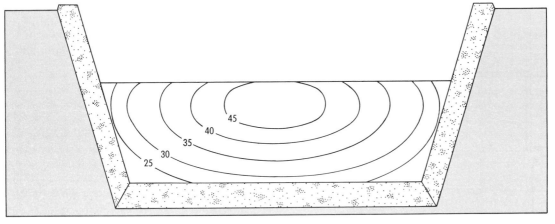

◀ Figure 5.12 Concrete channel

——— 40 ——— Isovel, showing the velocity of the water in cms/second

Natural channel

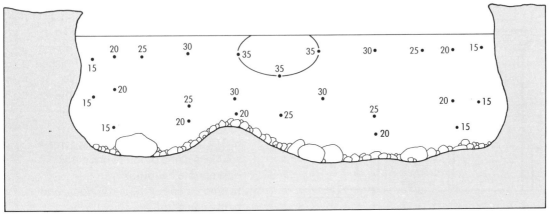

◀ Figure 5.13 Natural channel

• 30 Velocity of the water in cms/second

Erosion, transport and deposition in the river

River erosion

Look at the photograph and caption overleaf. The wearing away and removal of soil and rock debris by any natural process is called **erosion**. The river is capable of eroding its bed and banks in three ways.

● Abrasion

This is where the river picks up rocks and pebbles from the river bed, and flings them against the sides and bottom of the channel. This causes them to wear away, or abrade, the river channel. The process needs large amounts of energy and only takes place when the river is in flood. However, once abrasion has started, it can be very effective. For example, turbulent eddies, or whirlpools, in the river cause rapidly spinning pebbles to drill deep circular holes into the river bed, known as **potholes**.

● Hydraulic action

This is the tremendous force of river water on its own, smashing against the sides of the channel, and dislodging large quantities of material. Loose particles, such as sand and gravel, can be very quickly eroded in this way. Like abrasion, this process requires considerable amounts of energy, and only takes place when the river is in flood.

● Solution

This occurs all the time, and not just during a flood. Page 23 showed how rain water, and therefore river water, is slightly acidic. Water flowing along the channel constantly dissolves the surrounding rock, removing the material in solution. Some rocks, such as carboniferous limestone, are particularly vulnerable to this form of erosion. Solution is capable of removing large quantities of rock over a long period of time.

River transport

A river in flood possesses enough energy both to push and to carry huge quantities of soil and rock debris down the river channel. This is called river **transport**. A river will carry as much material as its energy will allow. The total amount of material being carried by the river is known as its **load**. This load can be transported in four ways.

● Traction

The largest boulders can only be moved during exceptionally violent floods. Then the force of rapidly flowing water pushes them along and they roll and slide along the bed of the river. As the large boulders roll along, they start to break up into smaller fragments. This is called **attrition**.

● Saltation

Fast-flowing water lifts up pebbles, sand and gravel and then drops them again so that they bounce along the bed of the river.

● Suspension

Floodwater churns up very fine particles of clay and silt and then carries them along within its turbulent current.

● Solution

Water flowing along the river channel is constantly dissolving the rock, transporting it away as a chemical solution.

River deposition

The river continues to transport its load until it loses energy, which happens as soon as the water starts to slow down. This might occur when the river reaches a less steep part of the valley, where it broadens out, or where it enters a lake or reservoir.

As the river loses energy, the first load to be deposited is the heavy and bulky material making up the **bedload**. The boulders and large stones which litter the bottom of many upland rivers are the bedload. The smaller sand and gravel particles travel much further, and are usually only deposited in quantity when the river is flowing much more slowly. Great deposits of sand and gravel give many lowland valleys their characteristic flat floors. The suspension load consists of fine clay particles. These are mainly deposited as mud at the bottom of the final stretches of the river, where it widens out into an estuary as it meets the sea. River deposited gravel, sand and silt is called **alluvium**. The solution load is never dropped. This constant input of dissolved rocks gives the sea its natural saltiness.

Exercise 5.6

Figure 5.14 tells you whether rock debris of different sizes will be eroded, transported, or deposited by a river flowing at different speeds. For example, coarse sand will be eroded by a fast flowing river (shown on the diagram at point A). It will be transported by a slow flowing river (point B), and deposited by a sluggish river (point C).

1   Working in small groups, use Figure 5.14 to complete Table 5.3.
2   a) Why do you think that very fine particles of silt are never deposited by the river, even when it is virtually stationary?
     b) Which is easier for the river to erode – silt or coarse sand? Give a reason for your answer.
3   State why each of the following people would find Figure 5.14 a useful diagram.

- An engineer strengthening the banks of a river which floods frequently.
- A hydrologist looking for the location of a new dam in an upland area.
- A farmer digging a new drainage ditch across his/her land.

▶ A river in Iceland running over deposits of alluvium. Look at the river erosion on the valley sides.

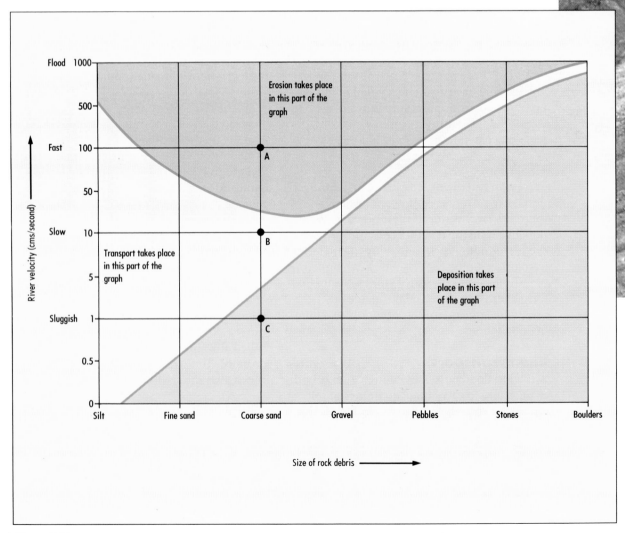

▲ Figure 5.14

Size of rock debris in the river	River velocity (cm/second)	Which will happen? Tick the right box		
		Erosion	Transport	Deposition
coarse sand	10			
gravel	1			
coarse sand	100			
stones	600			
fine sand	5			
pebbles	50			
silt	0.8			
stones	800			
boulders	900			
gravel	30			

◀ Table 5.3

Case study:
China's Sorrow – the Hwang Ho, or Yellow River

Area shown on the map

China

◀ **Figure 5.15** The Hwang Ho or Yellow River

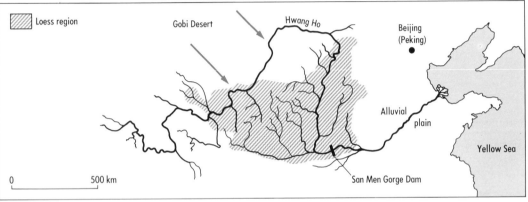

Loess region

Gobi Desert

Hwang Ho

Beijing (Peking)

Alluvial plain

Yellow Sea

0   500 km

San Men Gorge Dam

The Yellow River is located in north-east China (Figure 5.15). It transports vast quantities of material along its channel. Each year, it deposits 1.6 billion tonnes of sediment into the Yellow Sea. This is ten per cent of all the sediment brought down by the world's rivers.

Reasons for the sediment

The sediment in the Yellow River is loess. Loess is fine wind-blown soil. It is deep and fertile, and was blown out of the Gobi desert by strong north-westerly winds millions of years ago.

Seventy million people farm the loess region of China. Over the years, they have slowly stripped the area of its natural birch and pine forest to provide fuel and building materials. This, along with intense summer rain storms, has caused very rapid soil erosion. Vast quantities of soil are washed down steep-sided gullies and into streams which dump their load straight into the Yellow River. During the summer storms, some of these streams have to carry so much sediment that they flow more like thick porridge than water!

The land in the loess region is fairly steep. So the Yellow River flows rapidly through the area and transports the heavy sediment load away. However, as the river nears the Yellow Sea, the land flattens out, and the river begins to lose velocity. Huge amounts of sediment are deposited here, which over millions of years has created a vast flat alluvial plain. This rich farmland is the home for over 120 million Chinese.

Flooding on the Yellow River

The heavy summer storms in the loess region make the Yellow River a flood hazard. The people living on the alluvial plain have had to protect themselves by building flood banks along either side of the channel. However, the river is depositing its load so rapidly (one metre every 100 years) that it soon fills up the new channel, and threatens to spill over the top of the flood banks. The banks have had to be steadily raised, until now the river runs above the valley floor on a ten metre high embankment. Under these conditions, the collapse of a flood wall can be disastrous. The water pours out through the gap and down the embankment, ripping open the channel for many kilometres, and covering the land with floodwater and great depths of alluvium.

▲ Building up flood banks along the sides of the Yellow River

Controlling the Yellow River

The lower Yellow Valley can only be farmed intensively with the help of irrigation. One of the first irrigation dams was the San Men Gorge dam, completed in 1960. However, because of the huge amounts of sediment, the dam became half filled with alluvium in just ten years. The Chinese were forced to take drastic action. All the water is now emptied out of the dam just before the start of the summer storms. When the first floods arrive, the water which is carrying a lot of sediment is allowed to flow straight through. Only the later more sediment-free water is kept. This measure is now employed on all the Yellow River's dams.

The Chinese have made great efforts to control soil erosion in the loess region. They have terraced the valley sides and planted them with trees. They have built small check dams across every gully to hold back the sediment in tributary streams. Since 1981, China has taken tree planting especially seriously, and the loess region was singled out for special attention. By law every man, woman and child must plant between three and five trees per year.

China is one of the few countries in the world that has actually increased its forest cover.

Exercise 5.7

1   Which type of river erosion – abrasion or hydraulic action – is likely to be the most important in the loess region of the Yellow River? Give a reason for your answer.

2   Which type of river transport – traction, saltation, or suspension – is likely to be the most important as the Yellow River flows through the loess region? Give a reason for your answer.

3   Where is most of the alluvium deposited along the length of the Yellow River? Give a reason for your answer.

4   Describe the advantages and disadvantages of this alluvium to people living in the Yellow River valley.

5   Working in pairs, take a sheet of A4 paper and fold it down the middle. Use it to prepare an illustrated pamphlet explaining three methods which local people could use to control soil erosion in the loess region of China.

The long profile of a river

► Horseshoe Falls, Canada

◄ Figure 5.16 The long profile of a river

The **source** of a river is usually in an upland area, where rain falling onto the land forms small streams, which join up to form a river. The **mouth** of a river is where it releases all the water into the sea. Figure 5.16 shows how a typical river has a distinct concave shape when it is seen in section from source to mouth. This type of section is known as a **long profile**.

In the upper part of the river, the gradient of the valley is steep. The river does not contain a large amount of water, and runs in a small narrow channel. The channel contains large rocks and boulders which create a lot of friction, and tend to slow the river down. During floods, this debris rapidly erodes the river channel, cutting out narrow steep sided V-shaped valleys in the hillside.

In the lower part of the river, the gradient becomes more gentle. The river now contains more water from tributary streams, and so has a much wider and deeper channel. This cuts down on the amount of friction so the river often flows at a slightly greater speed.

After a flood, large quantities of silt, sand and gravel are deposited on the valley floor, giving the valley its characteristic flat floor and gentle sides. During normal flow, the river flows over these flood deposits in a series of wide bends known as **meanders**.

Exercise 5.8

On a copy of Table 5.4, write in the word 'increases' or 'decreases' in order to describe what happens to each of the listed characteristics as the river flows from its upper section to its lower section. For example, since the river becomes less steep in the lower section, the gradient decreases.

The smooth concave curve shown in Figure 5.16 rarely exists in reality. It is often broken by waterfalls or rapids.

River characteristic	Upper section to lower section
Gradient	decreases
River velocity	
River discharge (amount of water in the river)	
Width of river channel	
Depth of river channel	
Friction	
Amount of erosion	
Amount of deposition	
Size of bed load	
Steepness of valley sides	
Width of valley floor	

◄ Table 5.4

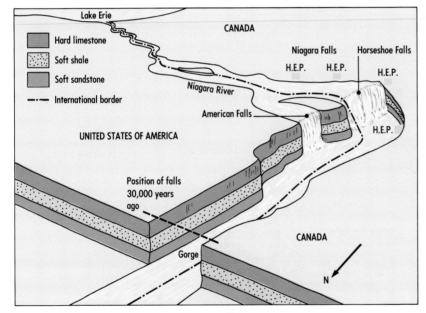

1 Debris swirls around in the plunge pool, causing abrasion of the softer rock

2 The hard rock is severely undercut

3 The hard rock collapses. The debris falls into the plunge pool

4 The debris swirls around in the plunge pool, causing abrasion of the softer rock. In this way, the waterfall eats its way upstream, leaving behind a steep sided gorge

☐ Soft rock   ☐ Hard rock

▲ **Figure 5.17** How a waterfall cuts back into the land

Waterfalls

Rivers often flow across several types of rock, some of which are more easily eroded than others. In the upper section of the valley, where the river is cutting down into the landscape, it cuts down faster into the softer rocks. Figure 5.17 shows how this forms a step in the river bed known as a waterfall.

Niagara Falls are one of the most spectacular waterfalls in the world. They lie on the Niagara River, which forms part of the border between Canada and the United States. At the falls, the Niagara River plunges over a 50 metre high cliff of limestone rock. At the bottom of the cliff a 50 metre deep plunge pool has been carved out of the soft sandstone and shale which underlies the limestone. The falls are eating into the cliff at the rate of one metre per year, and have carved out a gorge 11 kilometres long (Figure 5.18).

Rapids

Sometimes the change from one rock type to another is not enough to form a waterfall but it does create an uneven river bed. Figure 5.19 shows how this can happen when the water cascades over hard rock bands to form a zone of turbulent water known as a rapid.

Hydro-electric power stations, which use water to produce electricity, are sometimes built by waterfalls or rapids. Water is piped off at the higher level, and then dropped down under tremendous pressure on to a turbine at the lower level. The turbine drives a generator that produces the electricity. There are four power stations at Niagara Falls. They use 59 per cent of the river flow. This has dramatically slowed down the rate of erosion at the waterfalls themselves.

▲ **Figure 5.18** Niagara Falls

▶ **Figure 5.19** Bands of hard rock forming a series of rapids

Meanders

Look at the photograph, which shows a series of meanders on the River Add. A meander is a loop-like bend in the river, typically found best developed in the lower section of the valley. As the river flows around the bend, the fastest water tends to flow on the outside (Figure 5.20). During floods, the fast flowing water erodes the outer river bank, forming a steep **river cliff**. Deposition takes place in the slack water at the inside of the bend, to form a beach of alluvium called a **point bar**. The point bars build up over a long period of time to form a gentle **slip-off slope**.

It is not fully understood how a meander is formed. Some geographers think that it represents an efficient balance between the river's velocity and its gradient. During the early stages of formation, erosion at the outside of the meander causes it to grow in size. This makes the gradient of the river less steep, and the river starts to slow down. (The same principle is used by engineers building roads over very steep hills. They reduce the gradient of the road by meandering it through a series of hair-pin bends.) Eventually, a point is reached when the speed of the water is so slow that no further erosion can take place, and the meander stops growing. The size of the meander is now in balance with the amount of water flowing down the river.

This delicate balance between gradient and river velocity is easily upset. Since 1934, sixteen meanders on the River Mississippi have been artificially straightened, in order to shorten the journey time for barge traffic. The length of the river between Memphis and Baton Rouge has now been reduced by over one third. But doing this has increased the velocity of the river. This has resulted in the erosion of flood defences, and the river now threatens to break through at Turnbulls Bend. It would then follow a new course across the

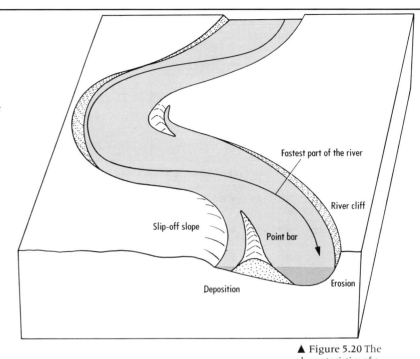

▲ Figure 5.20 The characteristics of a meander bend

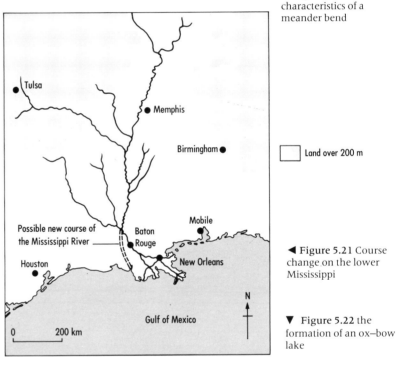

◄ Figure 5.21 Course change on the lower Mississippi

▼ Figure 5.22 the formation of an ox–bow lake

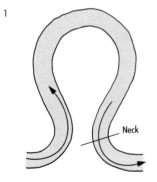

A very sharp meander bend, the two halves separated by a narrow neck of land

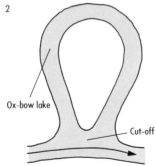

The narrow neck of land is quickly eroded away during a flood, leaving behind a loop of stagnant water, called an ox-bow lake

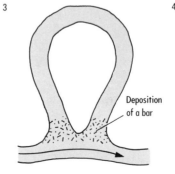

The river deposits a bar along the side of the channel, sealing off the ox-bow lake

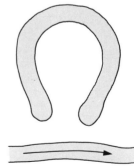

The ox-bow lake is slowly colonised by plants, and becomes land once more

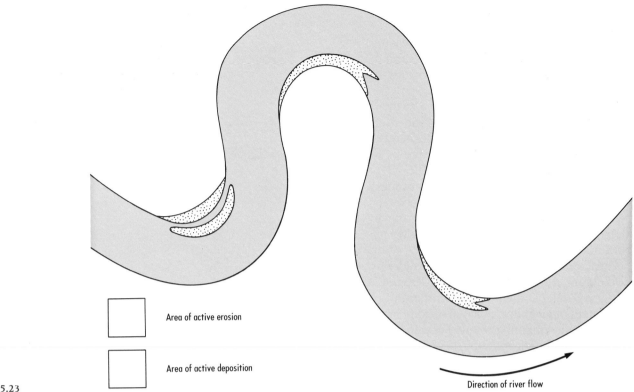

Area of active erosion

Area of active deposition

▶ Figure 5.23

Direction of river flow

▲ Meanders on the
River Add

delta and down to the Gulf of Mexico (Figure 5.21). This catastrophe would isolate the major industrial centres around New Orleans from the rest of the United States.

The balance between gradient and river velocity can also be upset naturally, for example during a flood. Figure 5.22 shows how the narrow neck of land on a sharp meander can be eroded away to create an **ox-bow lake**.

Exercise 5.9

1 On a copy of Figure 5.23, draw a long blue arrow to show where the fastest flowing water in the river is.
2 Colour the river bank red where active erosion is taking place, and brown where deposition is taking place. Complete the key.
3 Label one river cliff, one point bar, and one slip-off slope.
4 Look carefully at the photograph.
   a) 'The size of a meander is in balance with the amount of water flowing down the river.' Describe the evidence shown in the photograph that supports this statement, and explain how this balance occurs.
   b) Describe any evidence in the photograph that shows that the river channel is constantly changing.

Floodplains

The floodplain is an area of flat low-lying land on either side of the river channel. It is most developed in the middle and lower sections of the river.

The floodplain is cut out of the land by the movement of meanders. Figure 5.24 shows that as water sweeps round a meander bend, the point of maximum erosion is always slightly downstream from the **apex** (the centre of the bend). This means that meanders must always be moving steadily down the valley, as shown in Figure 5.25. As they move along, they erode a wide, flat floored valley bottom, known as a **floodplain**.

The floodplain is covered with deep deposits of alluvium. About 80 per cent of this is sand and gravel, left behind by the downstream movement of meanders as a trail of point bar deposits. The remaining twenty per cent is composed of fine silt and sand, deposited during times of flood.

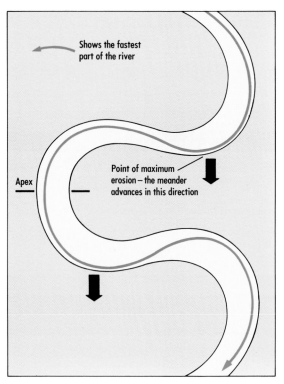

◀ Figure 5.24

Levées

When the river floods a valley, the fastest flow is over the submerged channel itself, where the water is deepest. (Figure 5.26) At the edge of the valley, however, shallower water moves more slowly. This means that the floodwater transports large quantities of sand and gravel along the submerged channel, but only fine silt at the valley edge. When the flood subsides, this material is deposited – the fine silt at the valley edge, and the larger sand and gravel particles on the river banks. In this way, repeated flooding can build up raised river banks call **levées**.

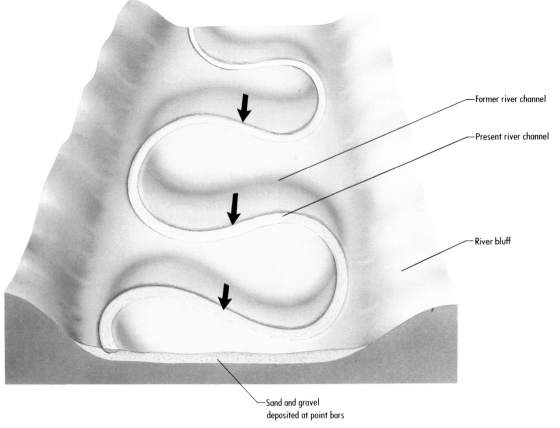

Former river channel

Present river channel

River bluff

Sand and gravel deposited at point bars

◀ Figure 5.25 The formation of a floodplain

► The floodplain of the River Thames at Marlow, Buckinghamshire, showing typical land uses

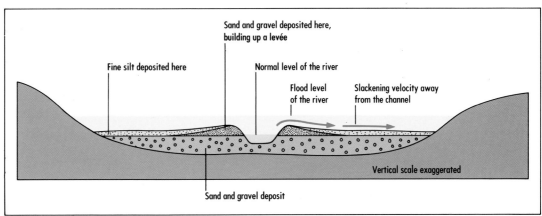

Fine silt deposited here

Sand and gravel deposited here, building up a levée

Normal level of the river

Flood level of the river

Slackening velocity away from the channel

Vertical scale exaggerated

Sand and gravel deposit

► Figure 5.26 Flood deposits causing a levée to form on the flood plain

Economic importance

Subsiding floodwater often becomes trapped behind the levées to form a region of poorly drained land known as a **backswamp**. The fine silt deposited here is very fertile, and makes the land good for agriculture. To be used successfully, however, it must be protected by flood banks, and artificially drained.

The sand and gravel deposited by point bars is also economically important. These deposits are easily quarried for use in the building industry. Once the material has been excavated, flooded gravel pits remain, which can be a dangerous eyesore. However, when landscaped, they can provide a recreation area for fishing, boating, and general parkland. They can also provide an important home for wildlife which would otherwise struggle to survive in a landscape which is increasingly cultivated or built over.

Braiding

Not all floodplains have meandering rivers. The river sometimes deposits so much material in its channel that it becomes choked with debris. It is forced to split up and thread its way through its own deposition. This is known as **braiding**.

Braiding often occurs when the river's load varies tremendously from season to season, such as in a river fed by glacier meltwater. Human activities can also cause braiding to take place. In the 1890s, Californian gold miners dug out vast quantities of material from the Sierra Nevada Mountains. The material was dumped into nearby rivers, which quickly choked and braided and they flooded land which had never been flooded before.

Exercise 5.10

1 Figure 5.27 shows the development of the River Thames floodplain around the village of Thorpe since 1920. Turn to an atlas map of London, and describe the location of this area.

2 a) Define the term 'floodplain'.
   b) Explain how a floodplain is formed.

3 a) Describe the way in which the River Thames floodplain was being used in 1920.
   b) Describe the way in which the land use had changed by 1950, and suggest a reason for this.
   c) Describe what had happened to the amount of land used by agriculture, the sand and gravel industry, communications and recreation in the area by 1980.

4 The Thorpe Park Leisure Complex is a major recreation facility developed out of 160 hectares of abandoned gravel pits by the Ready Mix Concrete group of companies. It contains a great range of leisure facilities including a water park, pavilions, special displays and showgrounds. It receives almost one million visitors per year.
   a) Explain why Thorpe Park is a good location for a large scale leisure complex.
   b) Describe the advantages and disadvantages of this development to people living in the village of Thorpe.

▼ Figure 5.27 Changing land use on the River Thames floodplain near Staines, south-west London

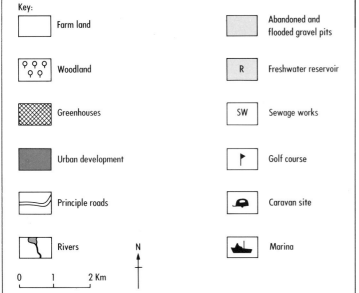

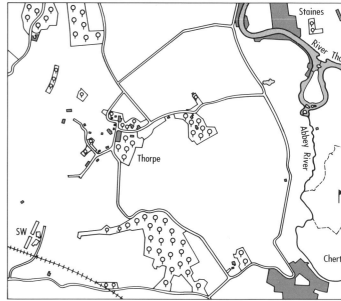

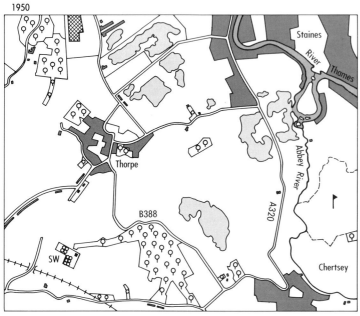

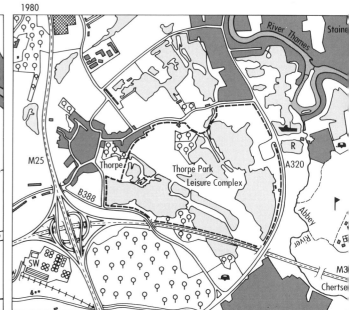

Terraces

A terrace is formed when a river is given more energy, and starts to cut down into its old floodplain to form a new one at a lower level – a process call **rejuvenation**. The terrace is left abandoned at the original level (Figure 5.28).

Rejuvenation occurs if the sea level falls, or if the land level rises. This could be due to either the tectonic forces discussed in Chapters 1 and 2; or to the effects of an Ice Age, which is discussed in Chapter 6.

Deltas

By the time a river reaches the end of its course, it is using all its extra energy to transport a full load of sediment. As the river enters the sea, it slows down and begins to deposit its load. Close to the shore, sand is deposited. Silt and fine clay is carried further out before it too is deposited. In this way, layers of different materials are built up on the sea floor. After thousands of years they form a huge platform of river sediment jutting out into the sea, called a **delta** (Figure 5.29). In the same way, smaller deltas form where a river flows into a lake.

A delta will only form under certain conditions. The river must have a high sediment load. The sea's tidal currents and waves must not be so strong that they remove the material faster than the river can deposit it.

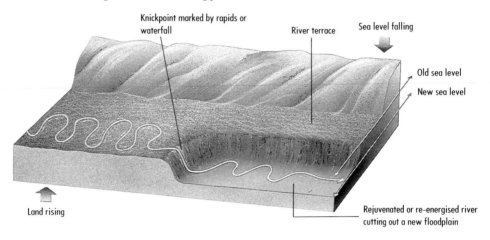

► Figure 5.28 The formation of a river terrace

► Figure 5.29 The formation of a delta

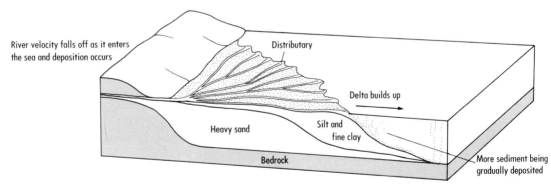

▼ Figure 5.30 Delta shapes

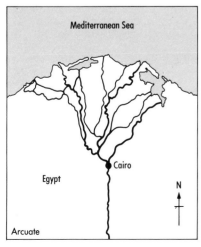

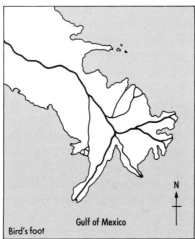

As the river enters the delta region, it is forced to flow over its own deposition. This causes the river to split up, or braid. Each of the resulting small channels is called a **distributary**.

There are two main types of delta, shown in Figure 5.30. When the river deposits its load uniformly over the whole area, the delta takes on a triangular shape, called an **arcuate delta**. When the river deposits its load rapidly, the river tends to push arms of deposition out into the sea, creating a **bird's foot delta**.

SIX

Ice sheets and glaciers

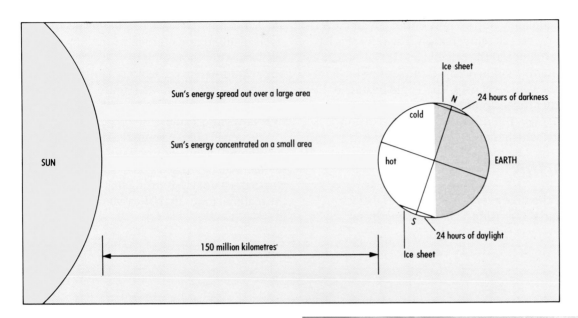

Sun's energy spread out over a large area

Sun's energy concentrated on a small area

SUN

150 million kilometres·

Ice sheet

N · 24 hours of darkness

cold

hot · **EARTH**

S

24 hours of daylight

Ice sheet

◄ **Figure 6.1** The position of the earth on 22 December, showing intense cold in the northern hemisphere

Eighteen thousand years ago Britain was covered by a huge sheet of ice. Most of the landscape of Britain looks as it does today because of the action of this ice. It was ice, for example, which carved out the dramatic mountain scenery of Scotland, Wales and the Lake District. At that time Britain must have looked rather like the scene in the photograph.

Ice sheets today

Seventy-five per cent of all the world's fresh water is still locked up in two great ice sheets – one in the north covering the Greenland continent and frozen Arctic Sea, and one in the south covering the Antarctic continent. Figure 6.1 shows how the main cause of these ice-sheets is the tilt of the earth's axis. Under polar conditions, temperatures vary between −10° and −50°C, and snow lies permanently on the ground.

Ice can also be found in the world's high mountain regions, where thin, or less dense, air causes very low temperatures throughout the year. Such regions are called Alpine environments.

How snow turns to ice

The intense cold of polar and alpine regions means that the amount of snow falling during winter is always greater than the amount of snow melting in summer. This is why there is always some snow lying on the ground.

Freshly-fallen snow is composed of fluffy ice crystals with lots of air spaces. This snow

▲ The Antarctic ice sheet – ice age Britain must have looked like this

has ten times the volume of the equivalent amount of water. Further snowfalls quickly compact (compress) the fresh snow into a much denser layer. Compacted snow that is at least one year old is called **firn**.

Firn takes another 25-40 years to turn into ice. This happens partly because the overlying snow compacts the snow crystals together, but the melting snow in summer is thought to be much more important. The meltwater percolates down into the lower layers of firn, and then refreezes around old snow crystals to form ice. In the Antarctic, where the snow does not melt, it can take up to 200 years for snow to turn to ice by compaction alone.

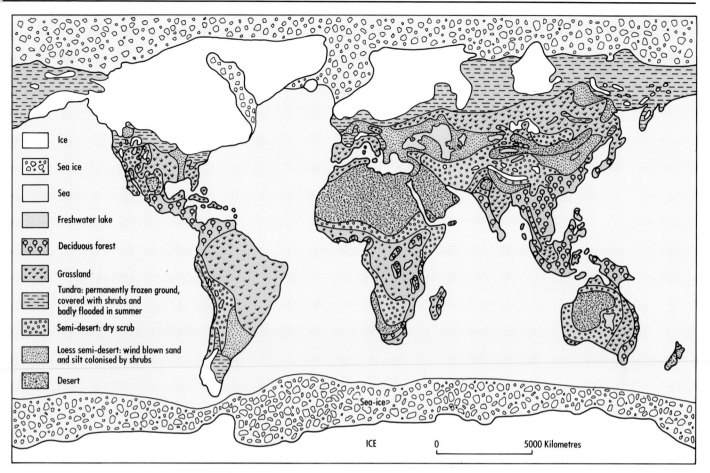

Key:

- Ice
- Sea ice
- Sea
- Freshwater lake
- Deciduous forest
- Grassland
- Tundra: permanently frozen ground, covered with shrubs and badly flooded in summer
- Semi-desert: dry scrub
- Loess semi-desert: wind blown sand and silt colonised by shrubs
- Desert

Sea-ice

ICE   0   5000 Kilometres

▲ Figure 6.2 The world during the last ice age, 18 000 years ago

Ice ages in the past

You have probably heard of **ice ages**. These are very long periods when the earth's average air temperature falls so low that the great polar ice sheets start to advance across the globe. The last ice age was the Pleistocene Ice Age, which began two million years ago, and ended just 10 000 years ago. Figure 6.2 shows a map of the world at that time.

Temperatures vary considerably during an ice age. The period of time when temperatures are falling is called a **glacial**. A period of warmer temperatures is called an **interglacial**. There have been a number of these temperature changes in the last 800 000 years. Some geographers think that we are still in a warm interglacial phase of the Pleistocene Ice Age, which has in fact not yet ended.

Geographers are still uncertain as to the cause of ice ages, but they may be the result of regular changes in the earth's orbit around the sun. As the distance from the earth to the sun increases, air temperatures become much lower. If this is correct, then the next ice age may start 23 000 years from now. World air temperatures have already fallen 2°C since 1300, and the period between 1500 and 1800 was a time of particularly severe winters, often called the Little Ice Age.

On the other hand, fossil fuel burning in industrial countries releases large quantities of carbon dioxide into the atmosphere. This gas blankets the earth, preventing heat from escaping – the so-called **greenhouse effect**. The resulting warming could delay the start of the next ice age, or even prevent it, perhaps causing the ice caps to melt and flood much low-lying land.

Exercise 6.1

Look at Figure 6.2 which shows a map of the world as it was during the last ice age, 18 000 years ago. Compare it with an atlas map of the world today.

1   a) Describe what happened to world sea level during the last ice age. Can you think of a reason for this?
   b) How did this change in sea level affect the geography of Europe?
2   Turn to an atlas map of world natural vegetation today.
   a) Compare the natural vegetation of southern Britain, France and Spain with that found 18 000 years ago.
   b) What happened to the tropical rain forest and desert regions during the last ice age?

The valley glacier system

Snow collects high up on mountain sides in Alpine areas. As the snow mounts up it turns to ice. Then gravity starts to pull it downhill. The ice tends to follow the easiest route, which is usually an old river valley. These long ribbons of slowly moving ice are called **valley glaciers**.

The speed of the glacier depends on how much ice collects in the mountain area and how much ice is melting away in the valley bottom. This glacier system is examined in exercise 6.2.

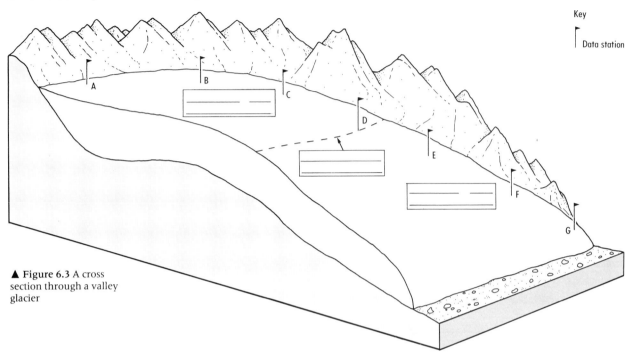

▲ **Figure 6.3** A cross section through a valley glacier

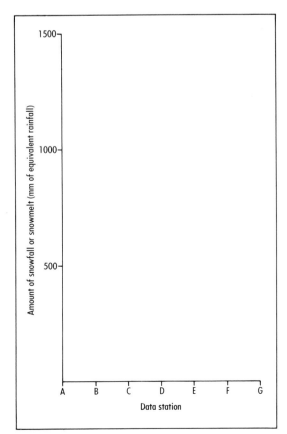

▲ **Figure 6.4**

Exercise 6.2

1 Look at Figure 6.3 which shows a section through a valley glacier. Table 6.1 shows the annual amount of snowfall and snowmelt at each of the seven data stations A–G set up along the length of the glacier.

a) On a copy of Figure 6.4 plot the snowfall at each data station, and join up the points with a blue line.

b) The fall of snow on to the glacier is called **accumulation**. Label your line 'accumulation'.

c) Plot the snowmelt at each data station, and join up the points with a red line.

d) The melting of snow is called **ablation**. Label your line 'ablation'.

e) At which data station does the amount of accumulation equal the amount of ablation? Label this point on your graph **'equilibrium point'**.

f) Colour in blue the area under the line on your graph where accumulation is greater than ablation.

g) Colour in red the area under the line on your graph where ablation is greater than accumulation.

Data Station	Amount of annual snowfall	Amount of annual snowmelt
A	1500	300
B	1300	500
C	1100	700
D	900	900
E	700	1100
F	500	1300
G	300	1500

Figures represent mm of equivalent rainfall

▲ Table 6.1

2   **a)** Now use your graph to complete the boxes on a copy of Figure 6.3 with the following captions – equilibrium line, zone of accumulation, zone of ablation.
b) The end of a glacier is called the **snout** – label this on your diagram.
c) The snout of the glacier lies at the end of the zone of ablation. What must the glacier ice be doing if the lower part of the glacier stays where it is, even though the ice is melting away?
d) Draw an arrow on your diagram, and label it 'direction of ice movement'.

3   **a)** Snowfall is not the only form of accumulation, and snowmelt is not the only form of ablation. Use the following list to complete the systems diagram shown in Figure 6.5: snowfall; snowmelt; avalanches on to the glacier surface; iceberg formation; direct evaporation of ice to water vapour.
b) Add a feedback loop to your diagram to show the effect of increased output on the size of the glacier store.

4   Look at Figure 6.6 which shows the relative amounts of accumulation and ablation during one year.
a) Describe what must happen to the glacier in January, and give a reason for your answer.
b) Describe what must happen to the glacier in June, and give a reason for your answer.

5   Copy out and complete the following passage, using the words provided below:

   The upper part of a valley glacier is called the ____. This is where snowfall is greater than ____, and where snow can survive long enough to become compacted into ____ and then ice. The lower part of a valley glacier is called the ____. This is where snowmelt is greater than ____. Between the two zones is a line where snowfall and snowmelt are equal. This is called the ____.
   As ice builds up in the zone of accumulation, ____ begins to pull it down the valley. However, the balance is not always the same throughout the year. During the ____ months, there is more accumulation than ablation, and the glacier ____ down the valley. During the ____ months, there is more ablation than accumulation, and the glacier ____ back up the valley.

   gravity
   advances
   zone of accumulation
   zone of ablation
   equilibrium line
   firn
   snowmelt
   winter
   retreats
   snowfall
   summer

▼ Figure 6.5

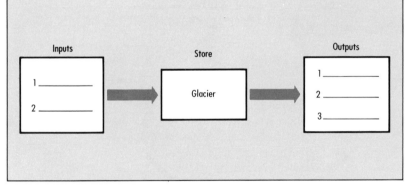

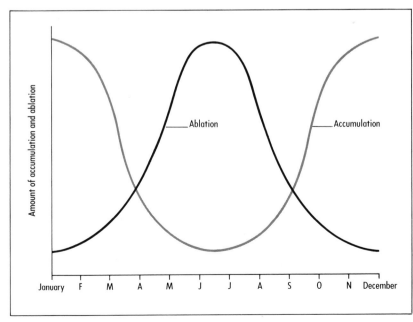

▲ Figure 6.6

Glacier flow

There are two ways in which glacier ice can move, depending upon the temperature of the ice:

Creep

This is important in very cold glaciers, such as those found at the edge of the Greenland and Antarctic ice sheets. Ice cannot flow like water because it is too brittle. However, 60 metres down inside the ice, there is sufficient pressure to make the ice move like soft plastic. The individual ice crystals arrange themselves into horizontal layers, and then slip over one another like a pack of cards (Figure 6.7). The speed of creep is about four metres per year.

Basal slipping

This is important in glaciers where the ice is only a few degrees below freezing, such as those found in Alpine areas. It occurs when the glacier slides over a thin film of meltwater under the ice. The speed is much faster than creep, and is about 50 metres per year.

As the glacier moves down the valley the ice is sometimes put under stress. This can happen when the glacier bends over a rock step in the valley floor, or when it spreads out as the valley widens. If it cannot bend quickly enough, it splits. These huge splits in the ice are called **crevasses**. (Figure 6.8).

Under tremendous pressure ice crystals start to move ...

and rearrange themselves into horizontal layers ...

which can then slip over each other like a pack of cards.

◀ **Figure 6.7** Creep

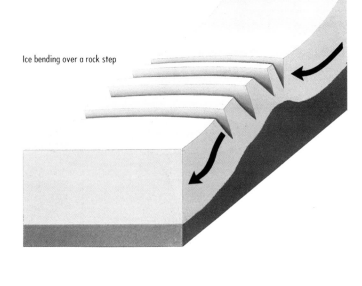

Ice bending over a rock step

Ice spreading out as the valley widens

◀ **Figure 6.8** Two causes of crevassing in a glacier

Valley glacier erosion

Glaciers can erode ten to twenty times faster than any river. Look at the deep valley carved out by the glacier shown in the photograph overleaf. This is due not only to the tremendous weight of the ice, but also to the fact that the glacier fills the whole valley, and not just a narrow channel.

Ice can erode in two distinct ways:

Abrasion

Rock fragments embedded in the ice grind away, or **abrade**, the bedrock. In one particular experiment in Iceland, a glacier abraded 25 millimetres of rock away from a marble slab in just three months.

Plucking

This occurs when ice melts and then refreezes around previously loosened rocks, pulling them out of the ground (Figure 6.9). **Pressure melting** is an important part of this process, where the tremendous weight of ice enables it to melt at temperatures a few degrees below 0°C.

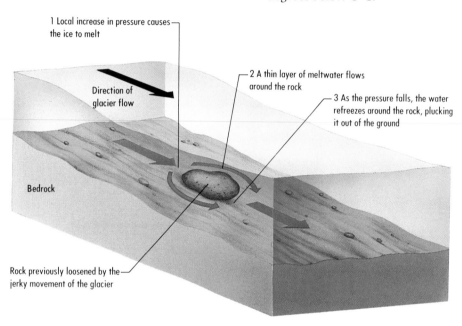

1 Local increase in pressure causes the ice to melt

Direction of glacier flow

2 A thin layer of meltwater flows around the rock

3 As the pressure falls, the water refreezes around the rock, plucking it out of the ground

Bedrock

Rock previously loosened by the jerky movement of the glacier

▶ **Figure 6.9** Plucking

Landscape features caused by glacial erosion

Nearly all glaciated valleys have areas of exposed bedrock which show evidence of abrasion and plucking. Abrasion mainly smooths and polishes the rock, although larger fragments of rock embedded in the ice can cause long scratches, known as **striations**. Plucking results in a steep rock step on the downstream side of any exposed rock. This can form a particular type of rock shape called a **roche moutonnée**, shown in Figure 6.10.

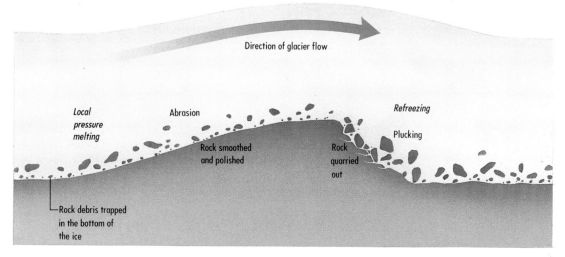

Direction of glacier flow

Local pressure melting

Abrasion

Refreezing

Plucking

Rock smoothed and polished

Rock quarried out

Rock debris trapped in the bottom of the ice

▶ **Figure 6.10** A cross section through a glacier showing the formation of a roche moutonnée

The most dramatic effect of ice erosion is the way in which the glacier changes the shape of the former river valley which you can see in Figure 6.11.

▼ **Figure 6.11** The effects of valley glacier erosion

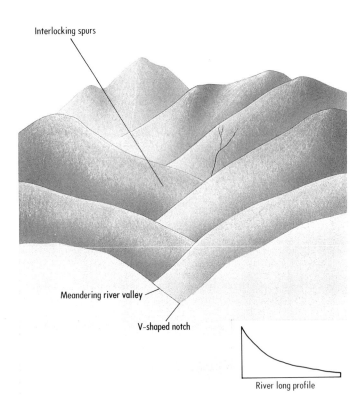

Interlocking spurs

Meandering river valley

V-shaped notch

River long profile

▲ (a) Before glaciation

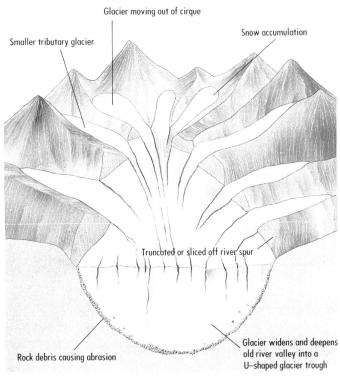

Glacier moving out of cirque

Snow accumulation

Smaller tributary glacier

Truncated or sliced off river spur

Rock debris causing abrasion

Glacier widens and deepens old river valley into a U–shaped glacier trough

▲ (b) During glaciation

Tributary glaciers do not erode as fast as the main glacier and so leave behind a hanging valley

Pyramidal peak

Cirque with tarn

Arete

Cirque

Waterfall

Frost shattering occurs on bare rock slopes

Scree slopes

Meltwater debris

Glacier long profile- the head of the valley becomes steeper and the floor is flattened out

Ribbon lake

Truncated spurs

Ribbon lake, filling a rock basin gouged out by the glacier

▲ (c) Just after glaciation

Recreation in the mountains – climbing, walking, skiing

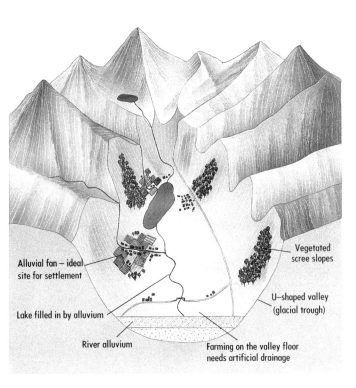

Alluvial fan – ideal site for settlement

Lake filled in by alluvium

River alluvium

Vegetated scree slopes

U-shaped valley (glacial trough)

Farming on the valley floor needs artificial drainage

▲ (d) After glaciation

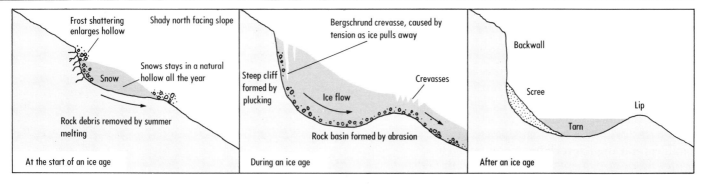

Frost shattering
enlarges hollow

Shady north facing slope

Snow

Snows stays in a natural
hollow all the year

Rock debris removed by summer
melting

At the start of an ice age

Bergschrund crevasse, caused by
tension as ice pulls away

Steep cliff
formed by
plucking

Ice flow

Crevasses

Rock basin formed by abrasion

During an ice age

Backwall

Scree

Tarn

Lip

After an ice age

▲ Figure 6.12 Stages in
the formation of a cirque

▼ An Alpine valley
glacier, showing ice-filled
cirques and arêtes

On the sides of many glacial valleys there are **cirques**, which mark areas where snow accumulated during the last ice age. Stages in the formation of a cirque can be seen in Figure 6.12. A particularly good example can be seen at Cwm Idwall, in Snowdonia, which is looked at on pages 76 and 77.

Often two neighbouring cirques erode back towards each other, leaving a knife-edged ridge between them. This ridge is called an **arête**. A particularly dramatic arête is Striding Edge on the slopes of Helvellyn in the Lake District. Sometimes a mountain can be surrounded by cirques on all sides, all eroding back into the rock until nothing is left but a central core. Such a feature is called a **pyramidal peak**. Perhaps the most famous pyramidal peak is the Matterhorn in the Alps, whose sheer sides have attracted climbers for many years.

Outside the valley, the glacier loses its dramatic erosive power as it spreads out and slows down. Nonetheless, its effect on the landscape can still be devastating. The slow march of the ice sheets across lowland North America and Europe two million years ago stripped away great depths of soil, polishing and smoothing the bedrock beneath. Many of these barren areas still remain, such as the knock-and-lochan landscape of north-west Scotland, or the vast Shield terrain of northern Canada.

Exercise 6.3

Complete the word net.

CLUES

1 After glaciation a river valley becomes one of these. (2 words)
2 Slice off a bend in an old river valley. (2 words)
3 A strange hollow, the birth place of a glacier.
4 A long thin lake.
5 Causes striations.
6 Fifty years after firn.
7 Tarn water has to escape over it.
8 Go near a backwall and you might fall over one of these.
9 Try and cut yourself on this knife-edge!
10 Yosemite – which valley got left behind?
11 To get this, first fill up your basin with water.
12 You do not find these peaks in Egypt.
13 Valley – after T but also after V.
14 No feathers after this sort of ice erosion.
15 Useless type of land for farming. (2 words)

Valley glacier deposition

Ice is capable of transporting enormous volumes of eroded material over considerable distances. Some of this material is carried along on top of the ice, as a **lateral** or **medial moraine** (Figure 6.13). The rest is carried within the ice itself. It is a jumbled mass of boulders and angular rock fragments mixed in with a heavy clay. As soon as the ice starts to melt the glacier deposits all this material. It is deposited throughout the whole zone of ablation. Once it is deposited it is called **till**.

A large amount of till is deposited directly on to the valley floor beneath the melting glacier. This layer of deposition is called **ground moraine**. It can be up to 30 metres deep (Figure 6.14) and includes large boulders called **erratics**. Erratics are rocks which have been transported a long way by the ice. They are a very different rock type from the rocks already in the surrounding landscape. Much of Denmark is made up of a jumbled mass of ground moraine, forming a peninsula jutting out into the North Sea.

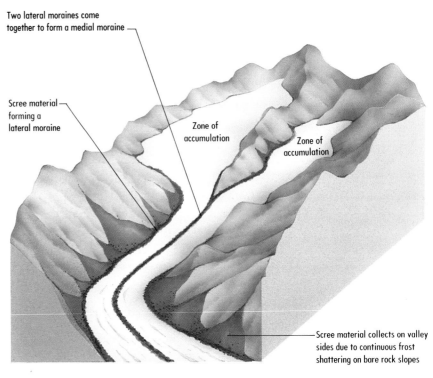

Two lateral moraines come together to form a medial moraine

Scree material forming a lateral moraine

Zone of accumulation

Zone of accumulation

Scree material collects on valley sides due to continuous frost shattering on bare rock slopes

▲ **Figure 6.13** The formation of lateral and medial moraines

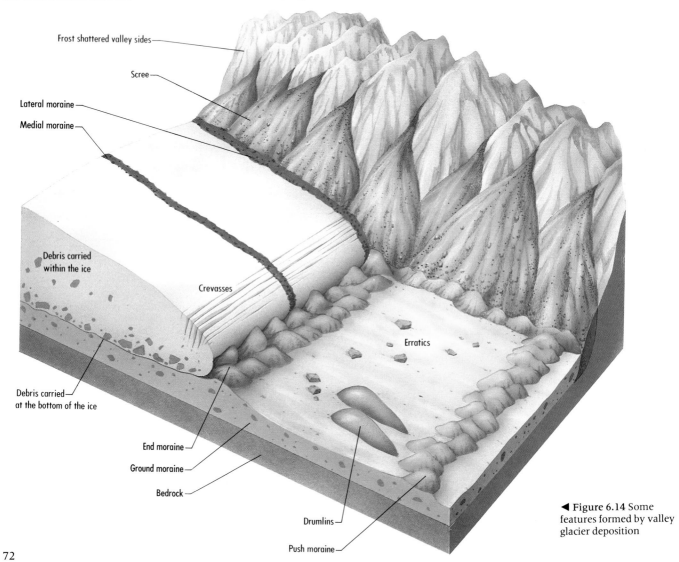

Frost shattered valley sides

Scree

Lateral moraine

Medial moraine

Debris carried within the ice

Crevasses

Debris carried at the bottom of the ice

Erratics

End moraine

Ground moraine

Bedrock

Drumlins

Push moraine

◄ **Figure 6.14** Some features formed by valley glacier deposition

Sometimes a glacier will re-advance over the ground moraine it has already deposited. When this happens, the ice moulds the till into small streamlined hills called **drumlins** (Figure 6.15). Most drumlins occur in large groups called **swarms**, and are typically about 35 metres high and 400 metres long. The end that faced the oncoming ice is blunt, the other end is long and tapering. So by looking at drumlins, we can get a good idea of the pattern of ice flow in an area. There is a particularly good example of a drumlin swarm in the upper Ribble Valley south of Settle, on the western slopes of the Pennines. Sometimes the ice moulds the ground moraine around a hard rocky outcrop, to form a **crag and tail** (Figure 6.16).

A great deal of till is not deposited by the melting ice as ground moraine, but is carried down to the glacier snout. Once there, it melts out to form a jumbled mass of material stretching across the valley floor, called an **end moraine**. Once the ice has retreated, the end moraine can often form a natural dam, creating ribbon lakes such as Wastwater in the Lake District which is 75 metres deep. Sometimes the glacier re-advances through the end moraine, bulldozing it along, and building up an even larger and more jumbled mass of till. This is called a **push moraine**.

► The Fox Glacier, New Zealand

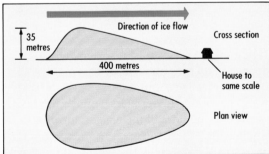

► **Figure 6.15** A typical drumlin shape

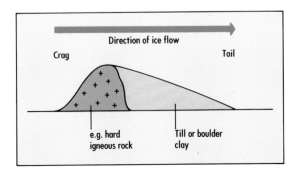

► **Figure 6.16** A typical crag and tail

Exercise 6.4

1   Define each of the following terms: till, end moraine, lateral moraine, ground moraine, push moraine, drumlin, erratic, crag and tail.

2   a) Lay tracing paper over the photograph, and make a simple sketch of the glacier, the mountain ridges and the meltwater river.
    b) How many of the following features can you arrow and label?
    zone of accumulation, zone of ablation, frost shattered valley sides, scree, crevasse, lateral moraine, meltwater river, meltwater deposition.
    c) Is this glacier advancing or retreating? Give a reason for your answer.

Meltwater

As the ice in a glacier melts in the summer months vast quantities of water are released. The water streams off the surface of the glacier in great torrents, pouring down crevasses and flowing under the ice to emerge at the snout as a river. The size and power of summer meltwater streams can be very great, but during the winter melting stops and the streams dry up.

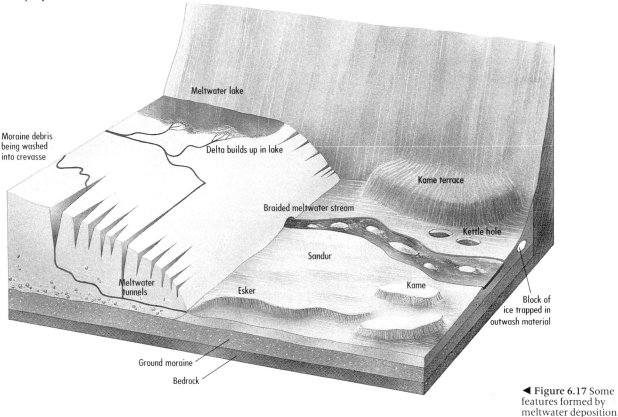

◀ Figure 6.17 Some features formed by meltwater deposition

Meltwater erosion

Because of its high velocity, **meltwater** can transport a heavy load of fine rock powder, sand and gravel. This material has come from glacier erosion, and enables the meltwater to erode channels under the ice called **meltwater channels**.

Meltwater can also collect in lakes at the sides of the glacier. If the water is high enough, it can spill over the watershed into the next valley, eroding an **overflow channel**. A good example of an overflow channel can be seen at Ironbridge Gorge in Shropshire.

Meltwater deposition

Meltwater flowing across the surface of the glacier washes rock debris from lateral and medial moraines into crevasses. (Figure 6.17)

When the ice melts, this unsorted debris is deposited on the valley floor. This mass of material is called a **kame**. The meltwater also washes moraine material into meltwater lakes trapped at the sides of the glacier, forming small deltas. These too create kames when the ice retreats. Eventually the whole lake becomes filled in with rock debris. When the ice melts, the debris is left as a shoulder of sand and gravel on the valley side, called a **kame terrace**.

Underneath the glacier, meltwater flowing in tunnels does not always erode. If it slows down it will begin to deposit material. When the ice melts, this deposition is left as a ridge meandering across the valley floor. This is called an **esker**.

As meltwater streams emerge from the

snout, they immediately begin to slow down and deposit their load – first large boulders, then gravel, and finally sand. Tremendous amounts of debris build up in front of the snout, forming a plain of deposition called a **sandur**. The meltwater streams braid as they flow over this area. Blocks of ice sometimes get trapped beneath the sandur. When these melt, a space is left and the material on top collapses into it. This leaves a hollow on the surface which then fills with water. These features are called **kettle holes**.

During the last ice age, great quantities of meltwater flowed across the land as the ice sheets retreated. When the meltwater was trapped, it formed vast inland lakes. In England, for example, one huge meltwater lake submerged much of the Midlands. This is known as Lake Harrison. Fine clay collected at the bottom of such lakes, to form nutrient-rich deposits, which have since weathered down into fertile soils.

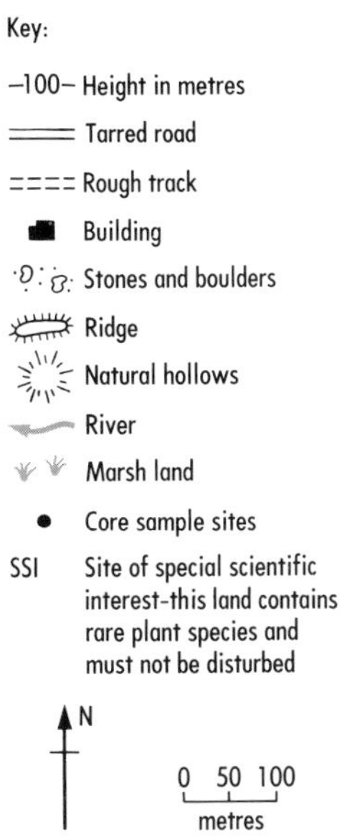

Key:
–100– Height in metres
Tarred road
Rough track
Building
Stones and boulders
Ridge
Natural hollows
River
Marsh land
Core sample sites
SSI
Site of special scientific interest-this land contains rare plant species and must not be disturbed
N
0 50 100
metres

200
175
150
125
100
Ridge Farm
Nitherton
Plot X
Plot Y
Plot Z
1
2
3
4
5
A
B
C
D
E
F
G
Dry Ridge
River Nith
SSI
100
125
150

▲ Figure 6.18

Key:
Topsoil
Sand
Gravel
Lake deposited clay
Till
Bedrock

Core sample site
1
2
3
4
5
0
20
40
60
Depth in metres
S
G
T
L

▲ Figure 6.19

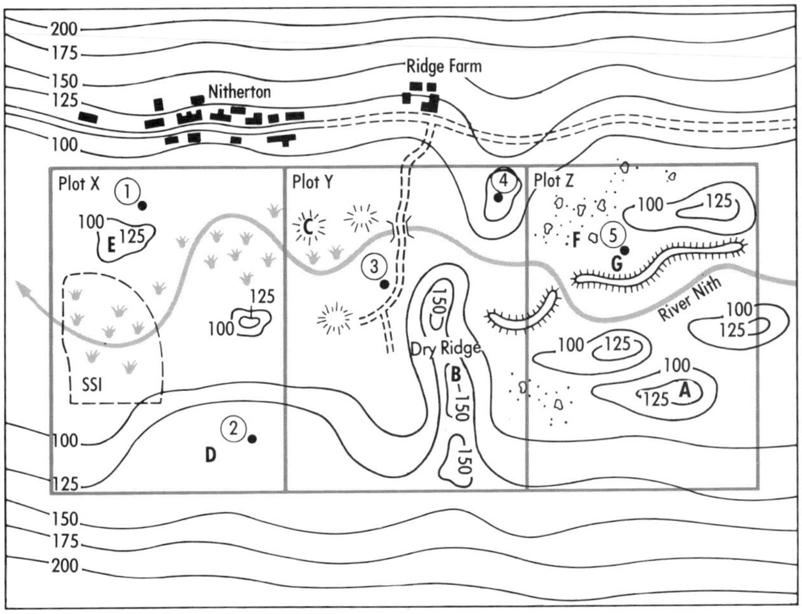

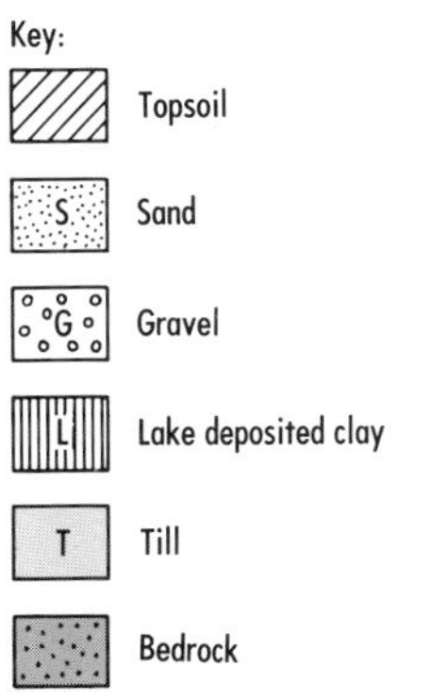

Exercise 6.5

Work in groups of two or three for this exercise.

You are geographers working for Glacial Surveys Limited. The Cumbrian Sand and Gravel Company has contacted you, and asked for a detailed survey of part of the Nith Valley shown in Figure 6.18. The three plots of land marked X, Y and Z are to be sold, and might contain useful sand and gravel deposits. The Company will extract any material by mechanical excavator, and transport it away by lorry.

1 a) Identify each of the following features marked A-G on Figure 6.18:
end moraine, drumlin, esker, kame, kame terrace, kettle hole, erratics.
b) Briefly describe each feature, and say how it was formed.
c) Three of these features will be of interest to the sand and gravel company. Which are they?

2 In order to take a closer look at the area, you have drilled into the ground and obtained core samples at the sites marked 1-5. The nature of these samples can be seen in Figure 6.19.
a) What is the most common deposit to be found on the valley floor to the west of Dry Ridge?
b) What geographical feature is this area most likely to be?
c) What is the most common deposit to be found on the valley floor to the east of Dry Ridge?
d) What geographical feature is this area most likely to be?

3 a) Write a detailed report to the Cumbrian Sand and Gravel Company. For each of the plots labelled X, Y and Z, you should:
1 state the physical features that can be found there,
2 describe the depths of sand and gravel that could be extracted by the company,
3 describe the amount of unwanted material that would have to be removed,
4 comment on any advantages and disadvantages for sand and gravel extraction on that plot of land.
b) Complete your report by writing a short paragraph suggesting which plot, or plots, of land the company should purchase for sand and gravel supplies, and why.

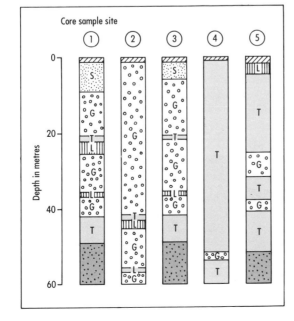

Case study:
The Nant Ffrancon
– a glaciated valley

The Nant Ffrancon valley is a typical glaciated valley located in North Wales (Figure 6.20). It is five kilometres long and over 500 metres deep.

Before the last ice age, the area was a high plateau with fast flowing streams and steep sided ravines. However, glaciers creeping out of the Snowdonia Mountains 15 000 years ago dramatically changed the landscape, creating deep glacial troughs.

Striations 500 metres up on the side of the valley indicate the probable depth of the ice. Squeezed between two high mountain peaks (Figure 6.20) the glacier

▲ The Nant Ffrancon in winter, looking north from the Rock Step at Rhaedr

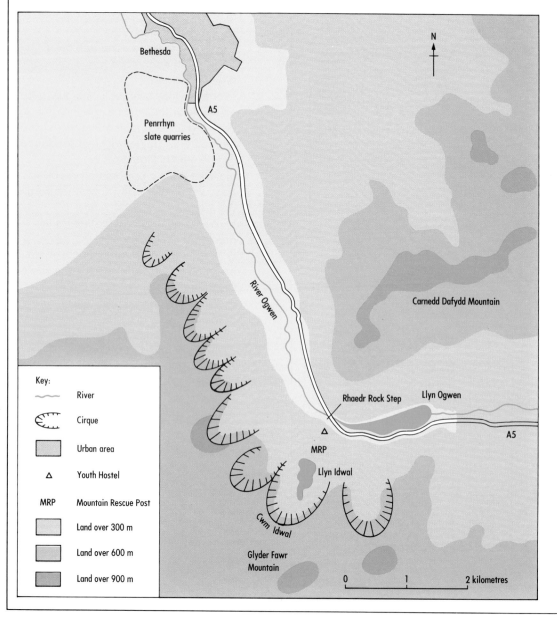

◀ Figure 6.20

speeded up through the Nant Ffrancon, overdeepening the valley to create a rock basin which is now filled with a ribbon lake – Llyn Ogwen. On the shady west side of the valley is a series of cirques, the most famous of which is Cwm Idwall, which contains a tarn. Ice flowing out of Cwm Idwall joined with the main valley glacier to produce a sudden increase in erosive power, forming a steep rock step at Rhaedr.

The valley floor is covered by a layer of ground moraine and a sandur, and is littered with erratics. The River Ogwen has since covered much of this with its own alluvium. It is poorly drained land, and will only support rough grazing for cattle and sheep. The steep valley sides have been frost-shattered, and the bottom of the slopes are covered by scree. Streams rushing down the valley sides have built up small alluvial fans. Rising above the floodplain of the River Ogwen, these make ideal sites for farms.

The valley has also provided an important route, or pass, through the mountainous Snowdonia region. The London to Holyhead road, now the A5, was built in the early 19th century. It is also an important tourist attraction, and there is a Youth Hostel and Mountain Rescue Post at Rhaedr.

Exercise 6.6

1 List the landforms found in the Nant Ffrancon Valley under the following types of processes:
glacier erosion, glacier deposition, meltwater action, weathering, river action.

2 Should we preserve the natural environment or develop it for our own use? Imagine that a planning application has been put forward, asking for permission to build a large leisure complex on the south side of Llyn Ogwen. The complex would include a 600 bed hotel, an outdoor leisure pursuits centre, a power and sail boat training centre, a children's adventure playground, and car parking for 1000 cars.

a) Divide the class into five groups, each group representing one of the interested parties shown in Table 6.2. Prepare your case, which will be made at a public enquiry in the Town Hall at Bethesda. You will need to elect a representative to speak on your behalf. Try to put yourself in the place of the person you are pretending to be, and ignore any views you might have.

b) Elect someone to chair the public enquiry. The chairperson should run the meeting in an orderly fashion, and then make a speech at the end, giving a reasoned account of why the application should, or should not, be given planning permission. Two others could help with this task.

c) Hold the public enquiry. Each representative should put forward his or her case, and then there should be time for a discussion of the points raised. The chairperson should then inform the meeting of his or her decision.

d) Imagine that you are a newspaper reporter covering the public enquiry. Write the front page of a local newspaper, describing what has happened, and include interviews with the interested parties.

Menai Property Group	Responsible for the planning application. They consider it to be a profitable development for their shareholders, and a sensible use of land which is of low agricultural value.
Bethesda Borough Council	Against the application – the area is a National Park, and even though the government has relaxed development restrictions for any ideas that generate employment, the proposal would destroy an area of great natural beauty.
Ogwen Tourist Association	Keen to see the development go ahead. It will provide jobs in an area of high unemployment, and could attract other tourist industries to the area.
National Farmers Union	Local farmers have already complained about the loss of expensive animals due to gates being left open and dangerous litter left lying around. The development would only increase the problem, and encourage trespass on their land.
Nature Conservancy Council	The area contains rare plant and animal species, as well as magnificent glacial scenery. We should always try to preserve our landscape for future generations to enjoy.

◀ Table 6.2 The interested parties

At the edge of the ice sheet

Many areas of the world lie on the edge of the great polar ice sheets. Although not covered by permanent ice and snow, these areas, which include arctic Canada, Scandanavia, and Russia are subjected to intense cold. They are known as **periglacial environments**.

The main characteristic of a periglacial environment is **permafrost**. Permafrost is permanently frozen ground. Very close to the ice sheet, where summer temperatures rarely rise above freezing, the ground remains frozen to a depth of 600 metres. Further away, the permafrost is usually thinner. During summer, the top few metres of permafrost usually melt. Large volumes of meltwater are released which cannot infiltrate into the still frozen ground beneath. The top layer of soil becomes saturated, and very mobile. This is known as the **active layer**.

Periglacial regions contain a number of distinct landforms. Over many years, due to the constant movement of the active layer, stones in the ground are forced to the surface. They are then sorted into very distinct patterns. This is known as **patterned ground**, as shown in Figure 6.21. Large earth mounds called **pingos** are also formed. One is shown in detail in Figure 6.22. Frost shattering is the most rapid weathering process on bare rock slopes, forming extensive scree. Strong winds blow off the nearby ice sheets, and transport fine silt over great distances, where it is then deposited as **loess**.

▲ Major engineering projects such as the BAM railway can seriously harm the periglacial environment

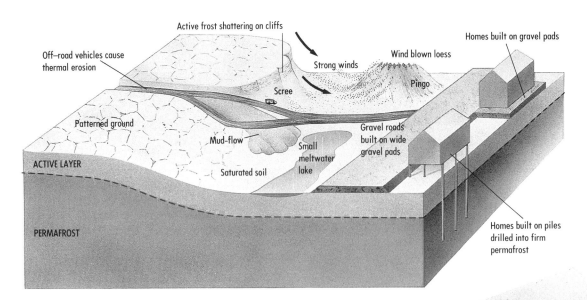

▲ **Figure 6.21** A typical periglacial landscape

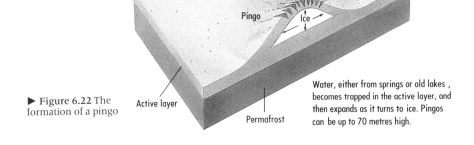

▶ **Figure 6.22** The formation of a pingo

Water, either from springs or old lakes, becomes trapped in the active layer, and then expands as it turns to ice. Pingos can be up to 70 metres high.

Engineering in periglacial environments

During this century, the growing demand for minerals and fossil fuels has brought industrial nations into the harsh periglacial environment. The active layer is an insecure foundation for buildings and roads. Houses have to be built on piles sunk into the hard permafrost beneath (Figure 6.21). Another solution is to use gravel pads, which act as insulators and prevent the active layer thawing out in the summer months. Without this protection, buildings would soon tip over as the ground melted, and roads would soon slip away.

As roads have penetrated these arctic wastelands, hunters and tourists have followed the mining engineers. Buildings and the use of modern off-road vehicles have increased dramatically, so that in many areas the overlying layers of moss and peat have been removed. This vegetation acts as an insulator, and without it the active layer melts too rapidly in summer. This has led to rapid soil erosion, and many mud flows. This process is called **thermal erosion**, and it has become a serious threat to many parts of the periglacial environment.

Exercise 6.7

1. Read the article 'A permanent way across the permafrost' and then answer the following questions:
   a) Name the resources that the BAM railway is designed to exploit.
   b) Why do you think that a second route had to be built to replace the existing Trans-Siberian railway?
   c) Why is Vostochny a key location in the BAM plan?

2. Many people consider the BAM project a threat to the Siberian environment. Describe some of the likely problems that large scale construction activity could cause.

3. A recent report stated:
   'The BAM railway has been built on a four metre high embankment, with the slope covered by rock fragments to reflect the sun,' and 'Many new Siberian houses have been built on stilts. When this has not been done, buildings heel over at crazy angles, often with window sills in the earth.' Explain the engineering problems and solutions behind these two statements.

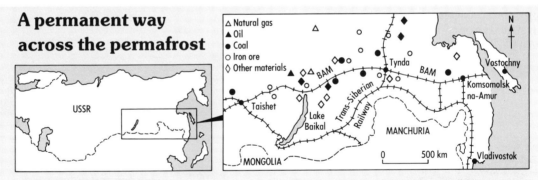

A permanent way across the permafrost

▶ **Figure 6.23** BAM picks its way through the minerals

The Soviet Union wants to bring civilisation to the permafrost of Eastern Siberia. It is one of the coldest places on Earth, with winter temperatures reaching −40°C, and the ground is permanently frozen to a depth of 1500 metres.

Soviet geologists have discovered deposits of fossil fuels and minerals in this region. The key to their exploitation is a project called BAM – the Baikal-Amur Mainline. This is a second Trans-Siberian railway, and will enable eastern Siberia to become a source of cheap raw materials for both Soviet industry and the newly industrialised countries of the Far East. This is important for Soviet foreign relations. Engineers are already building the largest port in the country at Vostochny, on the Pacific coast. It will export timber, coal and metal ores to Japan, in return for a trade of container loads of manufactured goods bound for Western Europe.

The completion of the railway will be a trigger for a second wave of investment. Soviet planners have designated 1.5 million square kilometres (seven times the size of Great Britain) of land as a BAM zone of development for mining, forestry and farming. More than a million people already live in an area once only visited by hunters.

The plan for BAM is immense and far-sighted, but nobody knows if it is money well spent. BAM's planners hope to cut down much of the 100 million hectares of forest in the zone. Experience elsewhere in Siberia suggests that, once trees are felled, soils are soon destroyed. A recent report stated that 'the scale of destruction around the railway exceeds anything previously known.'

Adapted from an article by Fred Pearce, *New Scientist*, 1 November 1984

SEVEN

Coasts

Coasts are the most varied and rapidly changing of all the natural environments. Look at the photographs on this page and note as many differences between the two coastlines as you can.

Waves

To understand how coasts can be so different, we have to look at **waves**. Waves are caused by the friction of the wind blowing over the sea surface. The wind tugs at the surface of the water causing the wave shape to move. This is the same as shaking a piece of rope. The wave shape travels down the rope but the rope itself remains in the same place.

The height of the waves depends on the speed of the wind, how long it has been blowing, and the distance over which it has travelled. This distance is known as the **fetch**. An increase in any of these factors will cause the wave height to increase.

When the wave reaches the shallow water near the coast, it breaks. This is because the base of the wave is slowed down by the frictional drag of the seashore. The way the wave breaks determines whether it has a **destructive** or **constructive** effect on the seashore (Figure 7.1).

▲ Land's End, Cornwall

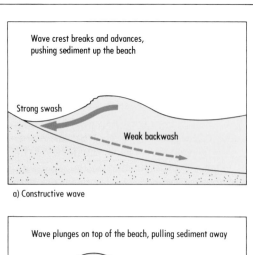

a) Constructive wave

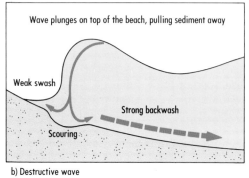

b) Destructive wave

◀ Figure 7.1

Exercise 7.1

1 Find Land's End in your atlas and match the following fetches and distances: westerly; south westerly; southerly; south easterly 200 km; 3500 km; 700 km; over 7000 km

2 Explain which wind direction is likely to produce the largest waves.

3 Read through the article on the Land's End disaster and answer the following questions:
   a) What type of waves were reaching the coast at the time of the accident?
   b) Using your knowledge of waves, explain why the cliffs at Land's End can be so dangerous.
   c) Describe one measure that could be taken to prevent such an accident occurring again.

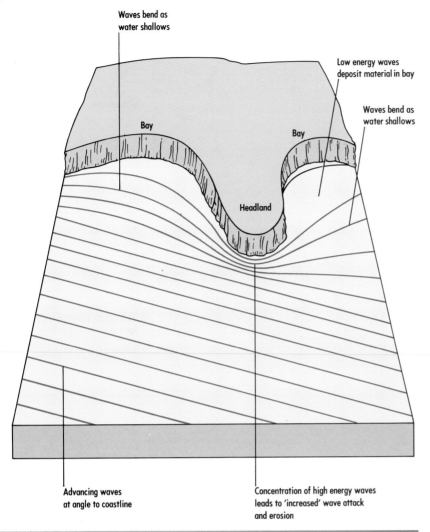

Waves bend as water shallows

Low energy waves deposit material in bay

Waves bend as water shallows

Bay

Bay

Headland

Advancing waves at angle to coastline

Concentration of high energy waves leads to 'increased' wave attack and erosion

The angle at which waves strike the coast varies considerably. The frictional drag of the sea floor causes waves to swing round until they are almost parallel to the shoreline. This bending of the waves is called **refraction**. With a very irregular coastline (Figure 7.2), wave refraction in the shallow water near the headland concentrates wave energy on the headland, so that the breakers there are much larger than those in the bays.

Waves out at sea are not just caused by the local wind. They interact with one another and can also be influenced by ocean currents and swells. **Ocean currents** are gigantic threads of water circulating around the globe. **Swells** are dying waves which are no longer influenced by the wind. Sometimes all these factors come together to form particularly large waves. For example, the seas off the south east coast of South Africa are notorious for huge killer waves. These waves can sometimes be over 30 metres high and are dangerous to shipping.

◄ Figure 7.2 Wave refraction

Four boys swept to sea at Land's End

FOUR BOYS were feared drowned last night after a school party had been caught by a large wave on boulders at Land's End, Cornwall. A girl was rescued from the sea by an RAF helicopter but the boys, aged between 10 and 12, were still missing when the coastguard called off the search at dusk.

The children were on a week's holiday at St Austell. Coastguards were called when a sightseer spotted a group of bedraggled children clambering out of the sea, which was running a heavy swell.

Police in Camborne said that six of the children who fell into the sea clambered out, but four boys were missing when teachers and police organised a head-count. The six were not detained in hospital.

The girl rescued by the helicopter was recovering last night in the West Cornwall hospital at Falmouth. She was believed to have been standing on a rock with one boy when the wave struck, while the three boys were on another rock nearby. The rescued girl suffered head injuries and hypothermia and was in a stable condition last night.

The pilot of the helicopter said: 'I think a teacher just plucked her from the sea as she fell in. There was a three or four foot westerly swell running which was making the water at the bottom of the cliff very treacherous.'

by Martin Wainwright
The Guardian, 7 May 1986

▲ Borth Spit Dyfed, Wales

Tides and tidal surges

Anyone who has spent the day at the beach will know about the tides. When the tide comes in, the beach is covered gradually with sea; when it goes out, the sea retreats. Tides are changes in the level of the sea. They are produced by the pull of the moon's gravity, and (to a lesser extent) the sun's gravity. When the sun and the moon are pulling in the same direction, very high tides occur. These are called **spring tides**.

The height of the tide varies from coast to coast, and different things affect it. In an estuary, for example, the water is constricted because of the narrower space it has to move in. So the tide is quite high. It can have a huge amount of energy which can be harnessed to produce electricity. Estuaries, such as the Severn, have been considered as possible locations for tidal hydro-electric power schemes.

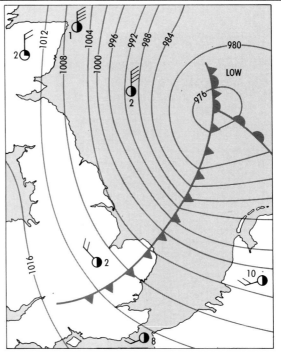

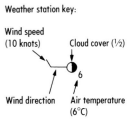

Weather station key:

◀ Figure 7.3 Synoptic weather chart for the North Sea at 01.00 hours, 1 February 1953

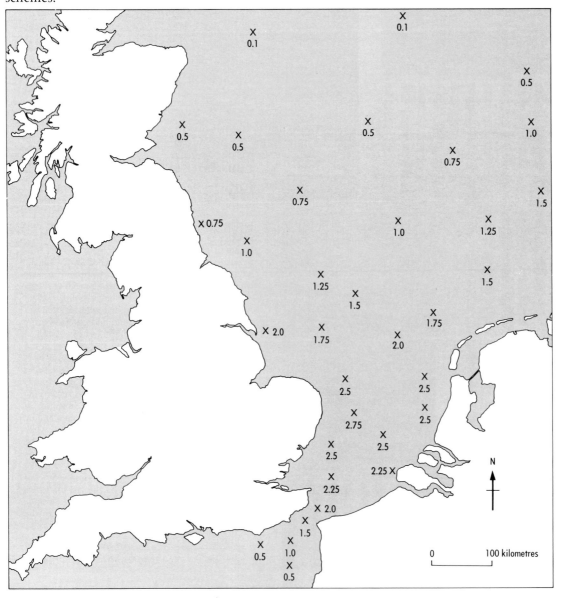

◀ Figure 7.4 Height of the North Sea and English Channel above normal spring tide state on 31 January 1953

▲ Canvey Island in the Thames Estuary, during the 1953 tidal surge

Tides can pose a serious threat to coastal areas. The North Sea is notorious for **tidal surges**. These are short-term increases in sea level. The most dangerous conditions occur when a high spring tide combines with low atmospheric pressure and strong winds. A low pressure system over the North Sea causes the sea level to rise and gale-force north-westerly winds push the water southwards towards the Straits of Dover. As you can see in Figure 7.3, this means that a great deal of water is trying to get through a very narrow channel. On the night of 31 January 1953 all these conditions occurred together in the North Sea. The result was a surge of over two metres. Since the wave heights were even greater, this led to extensive flooding along the east coast from the Humber to the Thames estuaries.
After this flood, coastal defences were built. One of the most expensive schemes was the completion of the Thames Flood Barrier.

Exercise 7.2

1   Study Figure 7.3 which is a weather chart for the North Sea.
    a) What was the direction of the wind along the east coast?
    b) What was the average wind speed in the North Sea?
    c) Given the prevailing weather conditions, briefly explain why the English and Dutch coasts were flooded.

2   **a)** Figure 7.4 shows the height of the North Sea above its normal spring tide state at various data stations. Complete this map by drawing isopleths at intervals of 0.5 metres from 0.5 to 2.5 metres.
    b) Briefly describe and account for the pattern of the abnormally high tide.

Coastal erosion

The erosive power of waves is enormous. It is estimated that, even in summer, breaking Atlantic waves exert an average pressure of 3000 kilograms per square metre. However, erosion is not caused just by the pressure of waves. There are four distinct processes of erosion.

- **Hydraulic action** is the explosion of compressed air trapped in cracks and crevasses of cliffs by advancing waves.
- **Corrasion** occurs when rock particles are hurled by the waves against cliff surfaces.
- **Attrition** is the breakdown of rock particles as they hit cliff faces and each other.
- **Corrosion** is the chemical decomposition of rocks by sea water and is most effective on limestone rocks.

The most common erosional landform found along the coast is the **cliff**. The steepness of cliffs largely depends on the rock type and structure. Cliffs made from hard rock such as granite are generally steep, like the cliffs at

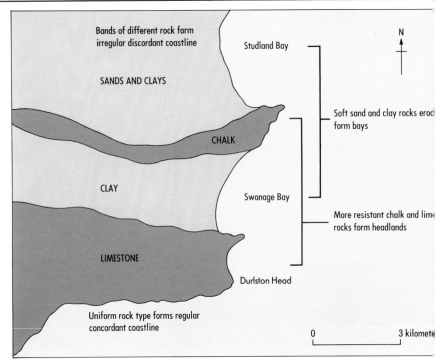

Bands of different rock form irregular discordant coastline

SANDS AND CLAYS

CHALK

CLAY

LIMESTONE

Studland Bay

Swanage Bay

Durlston Head

Soft sand and clay rocks erod form bays

More resistant chalk and lim rocks form headlands

Uniform rock type forms regular concordant coastline

N

0   3 kilomet

▲ **Figure 7.5** Contrasting coastlines of the Isle of Purbeck, Dorset

▼ The Normandy Cliffs

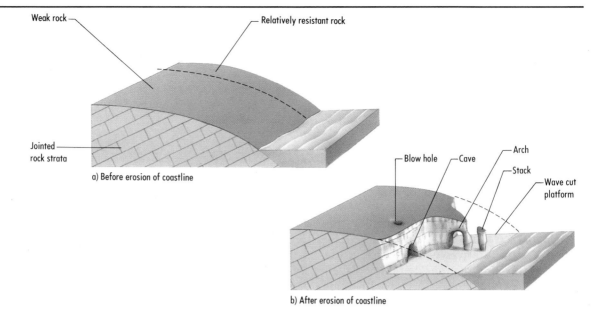

Weak rock — Relatively resistant rock

Jointed rock strata

a) Before erosion of coastline

Blow hole — Cave — Arch — Stack — Wave cut platform

b) After erosion of coastline

► Figure 7.6

Land's End shown in the photograph on page 80. These cliffs are also rugged, because the sea has attacked the rock along lines of weakness called **joints**. Cliffs made from soft rocks such as clay are usually gently sloping, unless they are being eroded at a fast rate.

The erosion of cliffs along weak spots such as joints and faults produces a distinct set of landforms (shown in Figure 7.6). Parts of a cliff may be hollowed out to form **caves**. Sometimes a section of the cave roof collapses producing a **blow hole**. Caves which occur in headlands or small rock promontories can be completely hollowed through to form **arches**. Eventually these are demolished by the sea, leaving behind isolated pillars standing away from the coast. These are called **stacks**. Both arches and stacks stand on **wave cut platforms** which are often exposed at low tide. These gently sloping surfaces are produced by wave attack of the cliff between the high and low water mark.

▼ Figure 7.7 Systems diagram of a cliff

Exercise 7.3

Study carefully the photograph of the Normandy cliffs.

1   Draw a simplified sketch map of the coastline and label the following features:
    a) headland,
    b) bay,
    c) arch,
    d) stack,
    e) dry valley,
    f) town of Etretat.

2   Explain why arches and stacks are common on this stretch of coastline.

3   Why do you think a large bay should have formed on this part of the coastline?

4   a) Look at exercise 5.1. Complete the systems diagram shown in Figure 7.7 by filling in the empty boxes with the following captions:
    Effect of gravity is increased.
    Waves transport debris away.
    Effect of rain water and sea spray.
    Weathered material stored on cliff.
    Wave attack erodes base of cliff.
    b) Use your completed diagram to explain how cliffs are formed.
    c) How would you modify this diagram to explain the steep cliffs found in Normandy?

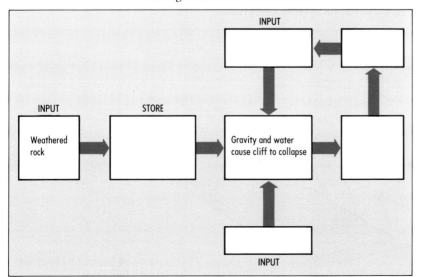

INPUT

INPUT   STORE

Weathered rock

Gravity and water cause cliff to collapse

INPUT

Case study:
The erosion of Barton Cliffs

Barton Cliffs are located in the middle of Christchurch Bay on the Hampshire coast. They are composed of horizontal layers of sand and clay. The cliffs are receding at a rate of one metre each year and several buildings once situated on top of them have already been destroyed.

Two processes are responsible for this erosion. The cliffs are often attacked by strong Atlantic waves which undercut the base of the cliffs and quickly remove fallen debris. Secondly, during prolonged periods of heavy rainfall, water seeps down to the clay and lubricates the junction between the sand and clay. This results in landslides.

▲ A housing estate threatened by the erosion of Barton Cliffs

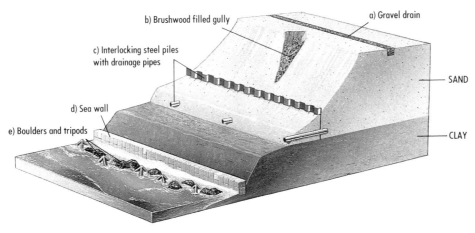

◀ Figure 7.8 Cliff protection measures at Barton on Sea

Exercise 7.4

Over £1 million has been spent on engineering measures designed to halt cliff retreat at Barton. Study Figure 7.8, which shows these measures, and explain how you think they are supposed to work.

Despite these measures, the cliffs are still retreating. In 1974, severe storms caused large sections of the cliff to slip and parts of the sea defences were destroyed.

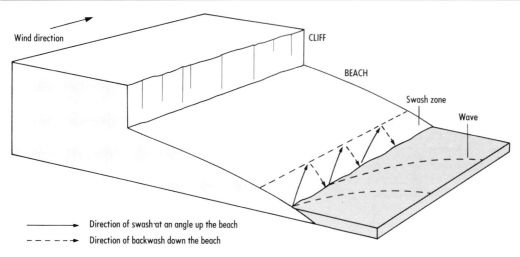

Direction of swash at an angle up the beach
Direction of backwash down the beach

◀ Figure 7.9 Longshore drift

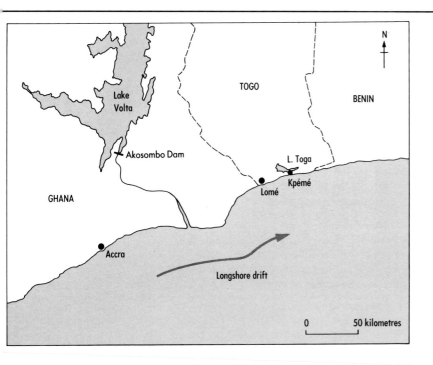

▲ Figure 7.10 Coastal erosion in West Africa

Spits, dunes, bars and tombolos

Spits often occur where there is an abrupt change in the shape of the coastline, for example, when it is broken by an estuary. The longshore drift tends to carry on in its original direction. The material is deposited, and gradually a long thin tongue of land is formed. This is the **spit**.

The end of the spit is usually hooked. This is because waves hitting the spit are travelling in a different direction from the longshore drift. Salt marshes develop in the area between the spit and the shore, and are produced by the deposition of fine sediment in the slack water. Salt resistant plants colonise these areas and help to trap more sediment.

Sand dunes are often found on spits and elsewhere. They are produced by wind carrying sand along the beach and depositing it in heaps around obstacles such as pieces of driftwood. These dunes grow as more sand is deposited by the wind. They are colonised by drought resistant plants such as marram grass. The grasses trap more sand and their long roots help to hold the dunes together, stabilising the area.

Sometimes material is deposited right across an inlet, resulting in a **bar** or **barrier** beach. Slapton Sands in Devon is a good example of this. Similar linear features can also be found off-shore. Probably Slapton Sands originated as an off-shore bar and storms pushed the bar towards the coast. Where a bar connects an island to the mainland it is known as a **tombolo**. An example of this is Chesil Beach in Dorset. All these landforms are shown in Figure 7.11.

Coastal deposition

The rock fragments torn away from cliffs by waves are broken down by **attrition** and transported along the coast where they are deposited as **beaches**. As waves break on beaches, the **swash** often pushes material up the beach at an angle. The **backwash**, however, moves directly down the beach slope. This zig-zag movement of material is called **longshore drift** and its direction depends on that of the wind (Figure 7.9).

Many local authorities have attempted to protect their beaches by stopping longshore drift. They construct wooden barriers called **groynes** which run across the beach at right-angles to the sea. These trap the sand and pebbles.

Sometimes such barriers have a bad effect on other parts of the coast. On the coast of West Africa, for example, longshore drift is from west to east. The building of the Akosombo Dam on the river Volta had already slowed down the rate at which sediment was carried into its estuary. However, more recently, a new port complex has been built at Lomé in Togo. This included a 1300 metre dyke out to sea. The dyke has stopped the longshore drift which means that although material is carried away from the coast, it is not replaced. The result is severe erosion to the east of the port. Already, part of the coast road has been swept away, and holiday villages and the phosphate wharf at Kpémé which is responsible for half of the country's export earnings are now threatened (Figure 7.10).

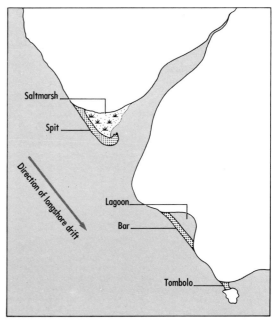

▲ Figure 7.11

Case study:
Deposition in the Dovey estuary

The estuary of the River Dovey is located in Cardigan Bay on the west coast of Wales. As with most of the estuaries in this bay, a spit has developed from the southern shore, extending northwards. The prevailing winds are south-westerlies and longshore drift is from south to north. The spit contains pebbles eroded from the Borth cliffs on the southern shore of the estuary. However, there is also material which is not local in origin. During the last glaciation, ice flows from Scotland probably deposited the pebbles of granite found in the spit, and ice flows from Ireland the pebbles of flint.

Sand dunes have developed on the northern part of the spit. Most of the sand comes from sand banks north-west of the dunes, and, to a lesser extent, from the beach. The main dune-forming winds are the dry north-westerlies which occur relatively infrequently. Behind the spit lies an extensive salt marsh, part of which has been reclaimed for agricultural use.

Exercise 7.5

1   Figure 7.12 is a map of the Dovey estuary at the end of the last ice age. The sea level is lower than it is today and the estuary has filled with sediment brought down by the river. Read the following information about the history of the spit, and on a copy of Figure 7.12 show the various developments that occurred from stage one to stage five.

1 – 7000 years ago
A 500 metre wide spit caused by longshore drift grows half way across the estuary in a northerly direction from Borth cliffs. Mark the bar on your map using the suggested symbol.

2 – 6000 years ago
The climate has become milder and behind the spit, the land has become dry enough for the development of a pine and oak forest. The bar has now grown three quarters of the way across the estuary. Mark both developments on your map using the suggested symbols.

3 – 5000 years ago
Sea level has risen and the climate has become wetter. The forest is flooded and decays. It is replaced by peat bog. Mark the peat bog in blue using the suggested symbol.

◀ The Dovey estuary, West Wales

4 – 4000 years ago

Sea level rises and successive storms push the spit one kilometre eastwards and beyond the western boundary of the peat bog. Mark and shade in a solid colour the new position of the bar.

5 – 2000 years ago to the present

The spit is now six kilometres long. On the northern section, approximately one square kilometre of sand dunes has developed. Immediately to the east of the dunes and extending in a north-easterly direction lies a kilometre wide zone of salt marsh. The area south of the salt marsh is reclaimed peat bog, except for a three square kilometre zone in the middle which is waterlogged. On the western side of the spit, tree stumps of the old forest are exposed at low tide.

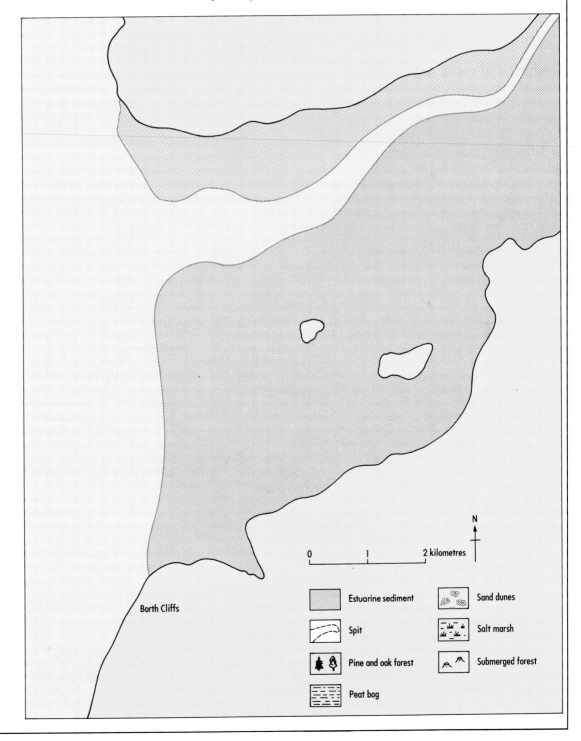

▶ Figure 7.12 Dovey estuary

Sea-level change

Long-term changes in sea-level are brought about by climatic change. This not only affects the type of processes, but also the rates at which they operate. Parts of today's coastal scenery, therefore, cannot be explained by present-day processes.

Evidence of falls in sea-level

The most striking evidence of a fall in sea-level are the flat platform surfaces along the coast, perched above the present sea-level. These features are known as **raised beaches** and are ancient wave cut platforms, some of which still contain beach pebbles.

Evidence of rises in sea-level

When the sea rises, it floods estuaries (see Figure 7.13). The south Devon coast contains a number of drowned river valleys which are known as **rias**. Much deeper are the drowned glaciated valleys known as **fiords**. The sea lochs of Scotland are fiords.

Submerged forests are also evidence of a rise in sea level. A number of these forests can be found in Cardigan Bay. The tree stumps are preserved by acid from the peat bogs which developed in the estuaries when the ground became waterlogged.

The crowded coast

Throughout the world, urban, industrial and tourist development is consuming the coastlines.

In the United Kingdom, local authorities and other organisations have made attempts to protect our coasts. In 1970, for example, the Countryside Commission designated over 1000 kilometres of coastline in England and Wales as **Heritage Coast**. The National Trust owns almost as much, as part of its campaign to save unspoilt coasts. However, these protected coasts have come under increasing pressure. Fursey Island in Poole Harbour on the Dorset Heritage Coast is one of Europe's biggest onshore oilfields. British Petroleum is keen to develop this field, as production costs are only one tenth of those in the North Sea. The Dorset County Council supported BP's drilling application because of the company's good reputation for safeguarding the environment.

One of the most serious threats to coasts is pollution. In Europe the coasts around the Mediterranean Sea are the most polluted. The problem has become very serious and the European Community is encouraging member states to reduce pollution levels.

▼ Figure 7.13 Evidence of sea-level change

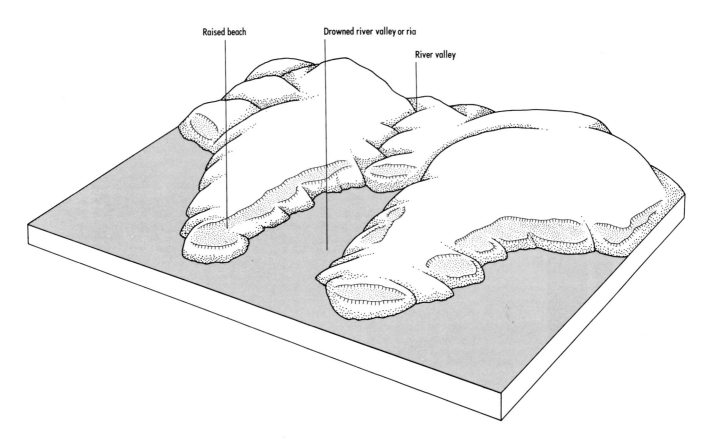

Raised beach

Drowned river valley or ria

River valley

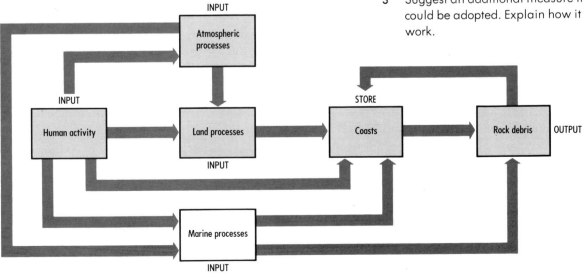

▲ Venice during normal high tide flooding in winter

Case study:
Venice – coastal mismanagement

Perched on the mud flats on the northern edge of the Adriatic Sea, the old city of Venice is under attack from both natural and human processes.

Large-scale reclamation projects for new factories have reduced the volume of lagoon. The removal of groundwater for industry at Marghera has caused a lot of subsidence. Deep water channels have also been cut in the lagoon to allow easy passage of large cargo vessels, and this has led to scouring of the lagoon bed which the city's foundations rest on.

The gradual melting of the Polar ice caps has resulted in a rise in sea-level, and this is aggravated by the high tides whipped up by the strong Sirocco wind, which effectively pushes water towards Venice. It was this combination of factors which led to the flood disaster of 1966.

Exercise 7.6

1   Make a copy of Figure 7.14 which is a systems diagram for coastal environments. Label some of the arrows using the information about Venice. Entitle your copy 'A systems diagram of Venice'.

2   Briefly explain how the construction of an aquaduct for Marghera and the strengthening of foundations will help to save the city.

3   Suggest an additional measure that could be adopted. Explain how it would work.

INPUT

Atmospheric processes

INPUT

Human activity

Land processes

INPUT

STORE

Coasts

Rock debris   OUTPUT

Marine processes

INPUT

▲ **Figure 7.14** Systems diagram for coastal environments

Deserts
The great arid lands

Most people think that deserts are great uninhabited regions of the world, baking hot and covered with sand dunes. This is a long way from the truth. Many deserts are in fact rocky and mountainous, and although temperatures can sometimes reach over 70°C during the day, they may fall as low as -10°C at night. It has also been estimated that 600 million people live on the edge of the world's deserts, thus making them far from uninhabited.

Deserts cover about 30 per cent of the earth's land surface. They are defined as being **arid** or dry regions, where the annual rainfall is less than 250 millimetres, and where evaporation is greater than rainfall. It is little wonder that deserts are considered to be hostile places!

Shaping the desert
Desert weathering

The desert landscape is formed by the weathering of bare rock, and then the erosion, transportation and deposition of this debris.

The dry nature of the desert climate means that weathering tends to be rather slow. One of the most important processes is **exfoliation** (page 22). During night-time cooling of the air, dew collects on the surface of bare rocks. Rapid evaporation of this water during the following day causes the growth of salt crystals. These force off the outer layers of the rock, which peel away like the skin of an onion.

This process tends to be most active where dew collects at the base of a rock. This can sometimes create mushroom shaped rocks called **pedestal rocks**.

Water action

Although there is little rainfall in a desert region, two factors make water the most important agent in the creation of the landscape. Firstly, many deserts were once much wetter than they are today, and many landscape features date from this time. Secondly, when rain does fall, it is often in intense cloudbursts, which cause destructive flash floods, and rapid erosion.

Irregular rainfall means that streams can only flow every so often. They are therefore called **intermittent streams**. When a storm does occur, however, the water rushes off the unprotected slopes as rapid overland flow.

▲ An oasis in the Sahara Desert, Algeria

This causes fast erosion, and creates steep sided gullies called **wadis** (Figure 8.1). These can fill so rapidly with water after a storm, that seasoned travellers avoid them.

The rainfall drains into large inland basins, called **playas**. As the stream slows down, it deposits its load, creating an **alluvial fan**. The water then collects in the playa for a short while as a **playa lake**. It soon evaporates away, however, leaving behind a mass of dried mud. After continual flooding and evaporation, evaporated salts sometimes collect on the playa surface. This is the origin of the Bonneville Salt Flats in Utah, USA. The perfectly flat surface of this playa, together with its accessibility, has made it an ideal location for attempts to break the World Land Speed Record.

▼ Figure 8.1 Desert landforms created by weathering and water action

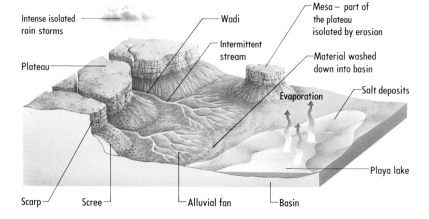

Intense isolated rain storms
Wadi
Mesa – part of the plateau isolated by erosion
Intermittent stream
Plateau
Material washed down into basin
Evaporation
Salt deposits
Scarp
Scree
Alluvial fan
Basin
Playa lake

Wind action

Wind action is particularly important in a desert region because of the fine nature of the weathered rock debris, and the lack of vegetation.

A strong wind is capable of picking up fine sand from the surface of the desert, creating a **sand storm**. This sand blasts bare rocky outcrops, forming them into long streamlined shapes called **yardangs**. The process of sandblasting can be very effective. Many of the explorers who crossed deserts by motor car in the early 20th century found paint soon stripped from their vehicles, and windscreens so badly scratched that they could not see out of them.

A more significant feature of wind erosion is the **deflation hollow**. The process of deflation is caused by wind blowing away loose dry sand. The resulting hollow tends to collect water, which causes increased weathering. The wind then blows the newly weathered rock debris out of the hollow, enlarging it still further. A particularly large deflation hollow is the Qattara Depression in the Sahara Desert – over 150 kilometres across, and almost 150 metres below sea level. Figure 8.2 shows how deflation hollows can also result in the appearance of an oasis.

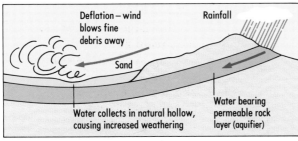

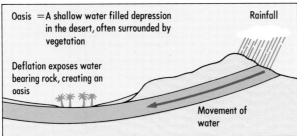

► Figure 8.2 How deflation hollows can create an oasis

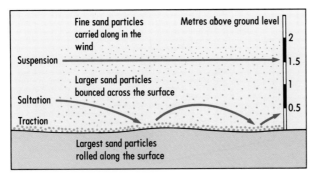

► Figure 8.3 Wind transport processes

Wind-deposited feature are called **dunes**. The wind transports the sand very much like a river, either by traction, saltation, or suspension (Figure 8.3).

As the sand grains are blown along, they collect into characteristic dune shapes, as shown in Figure 8.4. The familiar crescent shaped dune, called a **barchan dune**, is in fact quite rare. Sand dunes often collect together in one part of the desert to form a great sand sea, or **erg**.

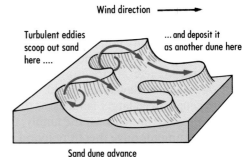

▲ Figure 8.4 Sand dune formation

Exercise 8.1

1 On a world map outline, using an atlas to help you, find and label the following deserts: Arabian, Atacama, Gibson, Gobi, Iranian, Kalahari, Mojave, Sahara, Thar, Turkestan.

2 **a)** Using the information in Table 8.1 copy and complete Figure 8.5 to form a pie chart showing the types of desert surface found within the Sahara.

   b) Look at the photograph and describe the different types of landscape that you would see if you visited the Sahara desert.

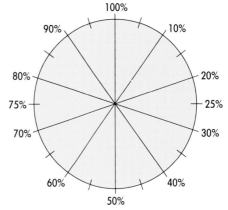

▲ Figure 8.5

Type of surface	Area of desert covered (%)
Mountainous and Rocky	45
Sand sea (ERG)	28
Salt crusted plain (PLAYA)	12
Desert pavement	10
Other	5

▲ Table 8.1 Types of desert surface found in the Sahara.

Desertification – the expanding deserts

All over the world, deserts are spreading into areas of previously productive land. This growing problem is called **desertification**. It is most serious on the edge of deserts in the developing world.

The causes

The main cause of desertification is a growing population on the desert fringe. Birth rates in many of these areas are four times that in the UK, and death rates have steadily declined due to improved medical care using knowledge from the developed world. Being poor, the people need larger families, as the more people who can work for the family, the better. Governments have therefore found it difficult to introduce birth control programmes. Figure 8.6 shows what happens next.

The solution

Some countries have attempted to solve the problem of desertification by using high technology. They build dams to store water, and then transfer it to the distressed areas for irrigation. In Egypt, for example, the Salhia Project uses Nile water to irrigate 23 000 hectares of land using overhead gantries. However, such projects often replace people with machines, and create unemployment in rural areas.

The low technology approach has therefore been more effective. Non-government organisations, such as charities and churches, have enlisted the help of the local people in countless small scale projects. The Gambia Christian Church, for example, has been digging local village wells, constructing dry season gardens, and instructing in soil conservation measures since 1975, with great success.

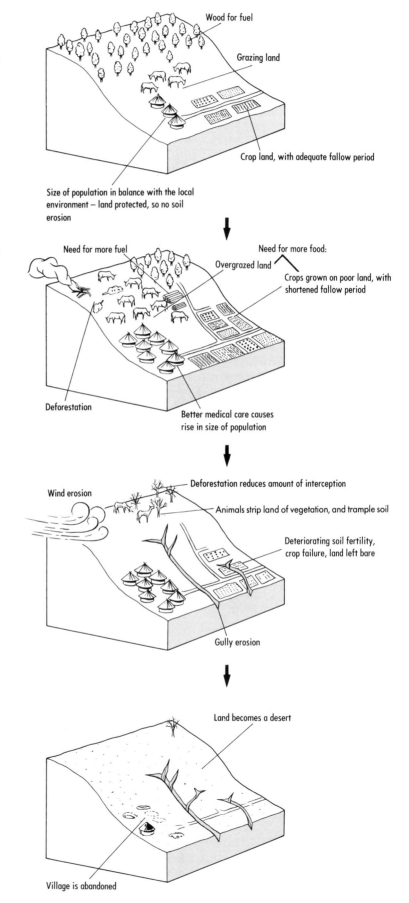

▶ Figure 8.6 The process of desertification

Case study:
The Sahel

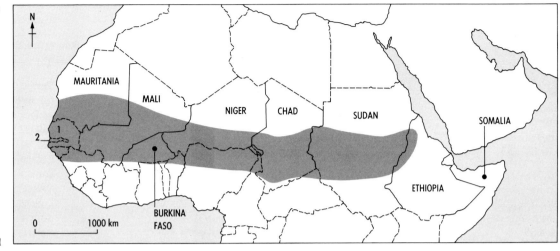

Approximate area of the Sahel zone

1 Senegal

2 Gambia

▶ Figure 8.7 The Sahel

Nowhere is desertification greater than in the vast area south of the Sahara called the **Sahel** (Figure 8.7). The desert is advancing south here at the rate of 6 kilometres per year.

Drought is common in the Sahel – there have been over twenty major ones in the last 500 years. In the 1960s, however, there was a period of regular rainfall. New grassland appeared close to the desert edge, and nomads migrated north to graze these new pastures, steadily increasing the size of their herds. Behind them, farmers also moved north, settling and cultivating land once used by the nomads. With prosperous agriculture and good medical care, the population rose by 22 per cent.

Then the rainfall became unreliable, and there has been a serious drought since 1970. The result has been a human and ecological catastrophe. The nomads soon found their new land overgrazed and trampled, with severe soil erosion setting in. The farmers found that their crops failed, leaving the land bare to the elements. Slowly, patch by patch, the land turned back into desert, and widespread famine set in.

Many people migrated away, looking for work in the towns and cities. Since the rural areas could not supply food to the urban areas, governments were forced to spend valuable foreign currency on buying food from abroad. This meant that there was little money to be spent on help for the drought-stricken areas.

With no economic output, the Sahel countries were unable to help themselves.

International aid for the region has become essential. Short term aid is aimed at stopping the famine. However, the distribution of food has been very difficult to organise. Lack of storage and port facilities means that 30 per cent of the food rots before it can be distributed, and in any case the worst hit areas are often far from any form of good road access. Making matters worse is the unstable political situation – local wars between neighbouring countries are common.

Long term aid for the region is more important. Money has been spent on digging village wells, stabilising sand dunes reforestation, improving road access, finding alternatives to firewood for fuel, and education programmes for better farming practices.

Exercise 8.2

1 a) Name the countries found within the Sahel region.
   b) List the problems that these countries face.
2 Explain why these countries find it difficult to expand food production under the following headings:
   climate, population growth, population movements, transport facilities, political strife.
3 Compile a scrapbook of cuttings from an informative daily newspaper showing examples of soil erosion and desertification taking place throughout the world.

A

Acid rain 24-25
Active layer 78
Alluvial fan 70, 77, 92
Arch 85
Arête 71

B

Bar 87
Beach
    formation of 87
    raised beach 90
Blow hole 85

C

Cave
    limestone 26-27
    sea 85
Cirque 71
Cliff
    Barton Cliffs 86
    formation of 84-85
Continental crust 8
Continental drift 9
Crag and tail 73
Crevasse 68

D

Delta 63
Deposition
    coastal 87-89
    desert 92-93
    glacier 72-73, 76-77
    meltwater 74-77
    periglacial 78
    river 51-55, 58-63
Drainage basin 28
Drumlin 73
Dunes
    coastal 87
    desert 93

E

Earthquake
    causes of 15
    effects of 15
    Guatemala City 6
    Mexico 14
Erosion
    coastal 84-86
    desert 92-93
    glacier 69-71, 76-77
    meltwater 74, 76-77
    periglacial 78
    river 50-55, 57-59
    thermal 79
    valley side 44-49

Erratic 72
Esker 74
Estuary (ria)
    Dovey estuary 88-89
    formation of 90

F

Fault
    rift valley 13
    San Andreas 11
    types of 13
Fetch 80
Fiord 90
Flooding
    Brent flood 36-37
    causes of 34-35
    tidal surge 82-83
    Venice 91
Floodplain
    formation of 60-61
    River Thames 62
Folding
    anticline 12
    fold mountain 10, 12
    syncline 12
    types of 12

G

Glacier
    flow 68
    system 66-67

H

Heritage Coast 90
Hydrological cycle
    controls on 32-33
    drainage basin 28
    flood hydrograph 34-37
    system 28, 30-31
    water supply 38-41
Hydrothermal activity 21

I

Ice
    ice ages 65
    ice sheets 64
    types of 64
Impermeable rocks 33
Intermittent streams 92

K

Kame 74
Kame terrace 74
Karst (see limestone scenery)
Kettle hole 75

L

Limestone scenery 26-27
Loess 54, 78
Longshore drift 87

M

Magma 16-17, 20
Mantle 8
Meander 58-59
Mid-Atlantic Ridge 9, 10
Mountains
    block 13
    fold 10, 12
Moraine, types of 72-73

O

Oasis 93
Ocean crust 8
Ocean current 81

P

Patterned ground 78
Periglacial activity 78-79
Permafrost 78
Permeable rocks 26-27, 33
Pingo 78
Plate boundary, types of 10-11
Playa 92
Pyramidal peak 71

R

Rapids 57
Refraction 81
Ria (see estuary)
River
    desert rivers 92
    discharge 35
    Hwang Ho (Yellow River) 54-55
    long profile 56-57
    processes 51-53
    velocity 50
Roche moutonée 69

S

Sahel 95
Sandur 75
Scree 22, 72, 78
Sea-level changes 90-91
Soil erosion
    desertification 94
    Nepal 48-49
    Sahel 95
    types of 44-47
Spit 87
Stack 85
Striation 69

Surging (tidal) 82-83
System
    drainage basin 30-31
    river valley 42-43
    sea cliff 85
    valley glacier 66-67

T

Terrace
    kame 74
    river 63
Tide 82
Till 72
Tombolo 87
Transport
    desert 92-93
    glacier 72
    meltwater 74
    periglacial 78
    river 51-55
    wave 87

U

U-shaped valley
    fiord 90
    formation of 70
    Nant Ffrancon Valley 76-77

V

Volcano
    Colombia 18
    Martinique 7
    types of 16-17, 20-21

W

Wadi 92
Waterfall 57
Water supply
    types of 38-39
    Kano Irrigation Project 40-41
Waves
    fetch 80
    refraction 81
    types of 80-81
Wave cut platform 85, 90
Weathering
    acid rain 24-25
    desert processes 92
    limestone (karst) scenery 26-27
    Taj Mahal 25
    types of 22-27